Your Backyard Homestead

A Step-By-Step Guide to Sustainable Living

Janet Wilson

ISBN: 978-1-951791-43-8

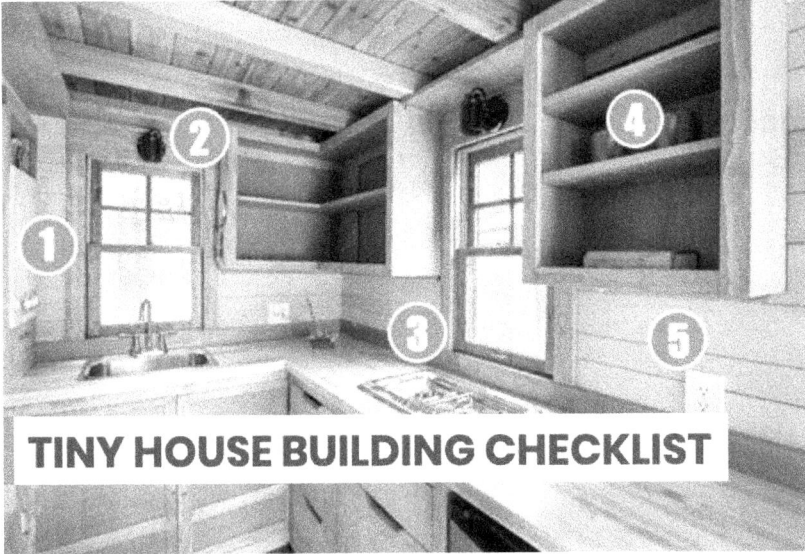

TINY HOUSE BUILDING CHECKLIST

Get Your Free Checklist:

- Learn How To Build Your Own Tiny House
- Includes Tiny House Plans
- Access to a Private Sustainable Living Community

Visit:
Janetwilson.org

Contents

Introduction

Due to the current world circumstances, more people than ever are turning to alternative means of providing their own food. You may think that your urban backyard is not the place for growing food; you may even think that it is impossible to grow enough food in a small backyard to make it worth your time. You couldn't be more mistaken.

Returning to your roots and providing your own food is an absolute possibility no matter where you live. There are methods that will work in any climate and you'll never need to worry about the food supply chain letting you down or the cost of food rising exponentially because of a lack of transport or a crop-killing bug or fire, or *any* natural disaster.

But, "I've only got a quarter of an acre yard here", you say. Well, that's just fine! On a small quarter-acre lot, you can grow up to *TEN-THOUSAND POUNDS of food!* That's right. It is possible to grow more than enough food for an average family on just one-quarter of an acre.

You probably won't grow that much in your first year or two, as you learn the ropes, but you can definitely grow enough to feed a small family without relying on the grocer for very much at all. Let's show you how to get started.

There are many reasons why you can benefit from growing your own food in the backyard. Some reasons you might not even have considered up until now.

1. How much do you spend at the grocery store now? A 600 squre foot garden (which is only 30 feet long by 20 feet wide) can yield enough food to save the average person or famly $600 per year at the grocery store. *Just imagine what your whole back yard can do!* A quarter-acre could produce as much as $10,000 dollars per year in food. You would have enough for yourself and sell some or trade to neighbors.

2. You'll know exactly what is in your food. You'll know how it was grown, prepared, and whether you can trust what is in the can because you are the one who canned it yourself, right there in your own kitchen. It leaves you with a deeper reverence for the planet as well; knowing and understanding how insects can be a nuisance while others are important partners in keeping a healthy garden. Learning how to be a steward of your own backyard garden will make you a more respectful resident of the planet.

3. Nothing is as empowering or uplifting as growing your own food. Getting children involved by giving them their own little square garden to be in charge of is a great way to make them enthusiastic about helping and eating their veggies too. Once children learn how to grow food, they are learning how to live and take care of themselves. It's arguably a lesson that all children should learn. It should be taught in school so that no one is ever completely dependent upon a system that is designed to provide you with food that isn't fresh, is full of steroids and additives, fertilizers and chemicals that many say are to blame for the rise in cancer and other chronic conditions that plague Americans.

4. You will never be at the mercy of anyone else for your food. It's empowering and gives you and your family a sense of security. Plus, your food will take on a new flavor because nothing beats the flavor of fresh food. There's a lot to be said for the mental connection that you have with food when you've planted it from seed and cared for it, harvested it, and preserved it yourself as well. The same goes for raising livestock. Many people claim that it is a grounding and spiritual process for them that reminds them of the connection we have to the earth and the food that it provides for us.

5. You could earn a side income from extra produce and selling young plants to people who don't want to sprout their own seedlings. It's a great way to earn some extra cash when people are all looking for some ways to bring in extra cash. Some people have successful YouTube channels just teaching others how to garden with many different methods and keep their plants healthy. There are outside of the box ideas from people all across the world and they are available to you on social media of all types.

6. Getting your hands dirty and creating plants; bringing something to life and caring for it, is soul-invigorating. Many people find it to be addictive and it turns into a hobby that they enjoy more than any other. Gardening can be a place of meditation and actually help you stay physically active, stretched, and calmer as you clear your mind and play in the dirt.

7. All summer long, you'll get your doses of Vitamin D that are gained from being outside in the sun. It's essential to proper

health and plays a vital role in mental health as well. It doesn't just end with the onset of fall either. You'll begin the hard work of preserving your harvest to ensure that it lasts the rest of the year until your next garden produces food again.

Chapter 1
Growing Food

Gardening: No One-Size-Fits-All Method

When it comes to gardening, the first thing that you should do is stop to really give serious thought to the things your family eats. Make a list of those vegetables and fruits. You should begin there. The next thing to bear in mind is what growing zone you live in. Knowing your zone will help you determine what will grow best in your location and when you need to plant your garden and start your seedlings.

The USDA has the best maps and explanations of growing zones and hardiness maps to help you determine the best vegetables to grow right where you are. Here's a link to the USDA website for more information and the map.

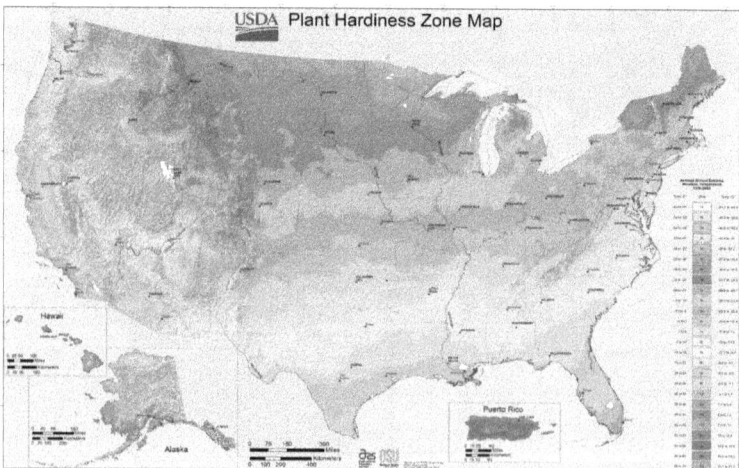

USDA Plant Hardiness Zone Map

There are also several types of gardening methods. Here is an overview of just a few of them:

Traditional Gardening - This refers to planting crops directly into the soil. You can do this in several different ways too. The main thing is that you have to have decent quality soil. If you don't have good soil, you can create good soil by mixing in purchased topsoil, fertilizer, creating and mixing in your own compost, and simply finding a way to get good nitrates into the soil.

Plants need nutrients, so get your soil in top condition and loaded with nutrients. Startin the fall, turn your soil over, add fertilizer, and compost. Plant a winter crop that you'll turn under in the spring. Within a few weeks, your soil will be in prime condition to plant your crops. There are many methods of traditional gardening also.

- **Planting in rows** - this is a method that is most often used by planting directly into the soil. It works well but, in some cases, it may not be the best method for smaller spaces.

  - **Three Sisters Gardening** - When you do plant in rows, consider using companion planting methods. This includes using the Three Sisters method of planting corn first, waiting a few weeks, and planting climbing beans that will ultimately climb the corn stalks while providing nitrogen to the soil, which the corn thrives upon. The last of the *sisters* to be planted is winter squash. It will thrive in the soil that has been conditioned and fed by the two prior crops and will provide you with a late-season crop. You can also plant tomatoes in three varieties together. This is a form of companion gardening. Here are some examples of vegetables that thrive when planted together:

○ What is **Companion Planting**? It's a method of growing foods that benefit each other. These plants either add something to the soil, provide a stalk for other plants to climb, provide shade to shade-loving plants, or provide a natural bug deterrent. It also prevents the wearing out of the garden soil that makes it necessary to rotate crops each year. It will naturally suppress weeds from growing also.

Some examples of Companion Gardening include:

- Dill, basil, and tomatoes planted together will protect the tomatoes from hornworms. Hornworms are very large types of caterpillars that you'll find clinging to your tomato plants. The first time you see one, you'll be horrified because they are as large as your thumb and they'll do a great deal of damage to your tomatoes.

- Sage planted near your cabbage plants will drive away cabbage moths. Cabbage moths can completely destroy your cabbage plants, chewing large holes in the leaves overnight. Your heads will look like Swiss cheese in very short order if you don't control the moths once they've found your cabbage.

- Marigolds planted around your garden will dissuade nematodes from taking up residence. These are very harmful insects to the garden. They also aren't favorites of rabbits or deer so will tend to keep some wildlife away from your garden as well. Now the thing is that there are good nematodes and bad nematodes. The good ones will help keep your soil aerated and loose, providing your garden with benefits. The harmful results of nematodes include eating the roots of your plants, killing them no matter what you do. Your best defense is keeping them away, to begin with, and marigolds are one way that doesn't involve potentially harmful chemicals.

- Carrots, parsley, dill, and turnips will attract the Praying Mantis which is a very good insect to have in the garden. He isn't interested in your garden plants but will feast on all the bugs he can, acting as a sentry over your delicate plants. Some people even buy these bugs so that they can hatch them and turn them loose in the garden each year. They are a wonderful learning experience if you have children as you'll have to hatch them indoors, in a special screened enclosure. You can have them shipped to you in a set, ready for you to hatch and eventually turn loose in the garden, where they will grow fat on the insects that they keep away from your plants.

- Spinach and other leafy greens will thrive in the shadow of corn, preferring the shade that is offered from this companion plant. The shade helps the delicate greens to not be burned by the direct heat from the sun and thrive in the humid environment at the base of the growing corn. The corn will also shelter tender greens from harsh winds.

- Catnip, rosemary, and sage planted with broccoli, cabbage, turnips, and kale will naturally keep pests away, especially the cabbage moth.

The Cabbage Moth as an adult moth.

- Zinnias planted near cauliflower will attract ladybugs. They're considered good fortune, but they'll also eat cabbage flies and other insects.

Zinnia with a honeybee.

- ○ Note that this is a wonderful way to incorporate herbs into your garden. You can benefit your plants by growing them and then use them to create the most amazing freshly made dishes in your kitchen. You can dry them and make your own seasonings. Imagine using your own dill to pickle your cucumbers each year or cooking your own spaghetti sauce and adding fresh rosemary, basil, oregano, and fresh garlic. *How rewarding that will be!*

- ○ **Square foot gardening** - This method is simply using square areas to plant crops. You may use companion planting with this type of garden. It can be formed using traditional gardening or with raised beds, which we'll also discuss later. Square foot gardening is a way that anyone can have access to reach crops easily. If you have mobility issues and use a wheelchair or a walker, it's difficult to get up and down garden rows.

Using the square foot garden method, you can create 4 squares, or six squares in a section, with a walkway all around that can be accessed in a wheelchair. This allows anyone to be able to grow a lovely garden. The yield can also be fantastic when you plant companion plants. Square foot gardening methods also are very easily adapted to doing companion gardening. You'll notice that many types of gardening can be combined into hybrid methodologies that allow you to personalize the type of garden that suits you the best.

○ **Vertical Gardening** - This is a fantastic method for those who have very little space. Even if you live in an apartment and only have a tiny patio or deck space, you can have a good amount of food growing in a vertical garden. The premise is simple -- when you can't spread out, spread *up*. You'll find that this method is a combination of two types of gardening -- container gardening and vertical gardening.

By using climbing structures, stacking pots, or even using old gutters mounted on boards, or PVC pipe and creating a hydroponic garden, which we will also go into more detail on, you can create amazing vertical gardens that will allow for as much food as you might grow in a large backyard. Honest.

○ **Container gardening** - This method is utilizing pots and any suitable container to plant your garden crops in. This allows them to be portable. If you are a seasonal resident, living in a camper for part of the year, this method allows you to take your crops with you, anywhere you go. It also can allow you to move things around when you need to when you've got a climate that makes it important to be able to bring plants inside.

Container gardening can be fun and even very whimsical in design. Allow yourself to think outside of the box a bit and you'll suddenly start to view what others may think of as junk as your next container

This is a prime example of thinking outside of the box. Note that the tires and engine have been removed, making the old car easier on the environment.

- **Hydroponic Gardening** - This is a method of gardening that produces and grows plants in a water-only environment, with no soil involved. While it seems a little strange, and there is admittedly a learning curve, it can be a wonderful way to grow food indoors, outdoors, and in situations where water is in short supply during the growing season. "Wait", you say, "how do you grow food in water with limited supplies of water?" Great question. It's a simple concept:

When you grow in soil, you must water your plants constantly. The soil pulls that water down, away from the plants very quickly. This means that a majority of water that you provide to the plant is actually pulled down into the water table, out of their reach before they can use it. By growing in water, you are providing them with water when and where they need it all of the time, without having to add much water because it is a sealed system. This actually allows you to use as much as one-third of the water that you'd use for the same plants grown in soil. An amazing way to conserve water!

This method can be used indoors by adding a UVB light over your plants so that you can grow food all year long, on decks, and in the yard. You'll need some special equipment to get started but it's a one-time investment and will keep your garden producing year-round in a greenhouse outdoors or inside with UVB lighting.

To get started, you'll need a large container with water, some small water lines, such as PEX tubing, a water pump, a filter, a timer, clay pebbles and grow baskets. You'll fill your grow baskets with a medium such as coconut husks, just to hold your seeds in place and give them something to root in.

You'll add a few clay pebbles on top of the growing medium to hold everything in place and stay wet, keeping everything moist on the surface. Your timer will kick your water pump on and off at intervals throughout the day so that water is pumped from below, up over the top of the plants where it will wet the growing medium and the clay pebbles, known as *Hydroton*.

Cut holes into a lid that will fit over your container, which could be as simple as a storage tote and lid combo. You'll cut the holes to match the size of your grow baskets, run your line from the water pump to the top of each plant (you'll need to get several t-fittings for your pipes to run separate off-shoots from the mainline).

Once your baskets are in place, your lines are run, your seeds are in the medium, they will eventually sprout. As they grow, the roots will reach further down into the water as it slowly evaporates. You want the water level to slowly evaporate to encourage the root growth for healthy plants. This means that once you've initially filled your container to just below the grow basket, you'll likely not need to add water for a very long time, if at all.

* An important note - once you fill your container, you'll want to add liquid fertilizer to it that is meant for hydroponic use. You'll also want to check the pH of the water after allowing it to circulate and run for two weeks prior to adding plants. This allows time for the chlorine to evaporate and the water to stabilize. A simple pH kit that you check pool water with will be fine. You'll want the range to be between 5.5 and 6.5 for optimum growth.

- **Aquaponics** - This is hydroponics on steroids. You're adding fish to the water tank underneath and by doing so, you are raising fish for meat and growing plants at the same time. We'll cover this in greater detail under the section *Small Space Livestock*. You may want to try hydroponics for a short while before you're ready to make the jump to aquaponics, but when you do you will wonder why you ever waited. You'll be able to grow vegetables and meat in the same space, which will require less intervention with fertilizers and provide you with hormone-free fruits, veggies, and meat.

The type of garden you choose will match your zone, your access to water, the type of soil that you have, and also say a lot about your style. A garden is a way to create and be imaginative. Let your personality shine through. Some people are practical gardeners, and everything is designed for ease and to be accessible.

Others will choose a garden design that makes them feel as though they have entered into another world. This is their getaway, the place they meditate and spend quality time inside their own minds. It will be tranquil and private. Which type of gardener are you?

Planting and Caring for Fruit Trees

When it comes to adding fruit to your garden space, you can choose the ground cover and bush-growing fruits. These will include tomatoes, strawberries, blueberries, raspberries, and blackberries most commonly. You can also consider a fruit tree.

Fruit is an important source of vitamin C and other vitamins and minerals that keep your body healthy. Fruit in your diet has been shown to improve immune system function, even helping to ward off the common cold.

The caveat with fruit trees is that they must grow for a while before bearing fruit. Miniature fruit trees will bear fruit faster and take less space. They're also easier to harvest as they don't grow especially tall. If you desire apples, for example, consider a miniature that is a 5-in-1 tree.

This is an apple tree that is grafted by taking stems from five different apple trees and forming one tree that will bear five different types of apples! Now you have apples for pie, apples for applesauce, apples for snacks, and apples for dehydrating. No matter who in your family has a favorite type of apple, you can grow it on one single tree. These types of trees are fast-growing and provide years of fruit for your family.

Make sure that you buy your fruit trees from a greenhouse or nursery that will guarantee that your tree is healthy from the start or they will provide you with a new start. Also, some types of fruit trees need to be bought in pairs so that they pollinate each other. Some species of apples must have a second tree to pollinate each other. Know which type you are buying and if you need two. If you don't, you'll have a pretty tree that never bears any fruit at all.

On the other hand, most varieties of peaches, nectarines, and cherry trees are self-pollinating and can produce fruit alone, as the only tree. Carefully research your fruit trees before buying. A fruit tree shouldn't be an impulse buy because they require care to do well and planning to ensure that you've gotten a good tree for your particular environment.

Some miniature trees will do well in pots that can be moved outside in the summer and brought indoors over the cold winter. Lemon trees are a great potted tree. Many people grow them in all sorts of climates. You'll simply need it near a window in the winter so that it can get sunlight while staying warm inside, basking in your care.

Properly transplanting a tree will make the difference between a tree that thrives or a tree that dies. Most nurseries provide you with a sheet of instructions for proper planting and it's important that you follow these instructions to the letter. Many nurseries will guarantee their new trees, but you'll need to have done things correctly. Some nurseries will plant the tree themselves to ensure that the tree is properly set in place, mulched, and watered to start its new life.

Ensure that you are planting the tree at the right time of the year, for starters. Early spring is typically the best time to plant your young fruit trees. You'll want to protect them from deer and other animals that will nibble on them, so place a wire cage around them to offer them this safety as they start setting their roots and growing to a size that is more resilient.

When choosing a spot for your trees, be certain that they won't grow into power lines or that you are digging down into water lines or underground cables. Make sure that there are no underground utilities buried on your property before you dig. Each state has a hotline and work crews will happily come out and mark the locations of underground cables so that you don't accidentally damage them.

Now that you've got your location, you'll need to dig a hole that is plenty big and gently slide your tree into the hole. Many trees are now packed in such a way that they've got all they need in their root ball, wrapped with materials that will compost into your soil. You may need to cut the bag to release the roots. Follow your directions that came with your tree. Backfill your soil to the hole and ensure that your tree is perfectly straight before doing so.

Mulch the base of your tree, water it well, protect it with a cage and fertilize it accordingly. Fruit trees need proper nutrition in order to grow healthy and disease-free. They will not bear fruit until they are old enough. This is another case for dwarf fruit trees. These will often bear fruit in their second year and grow very quickly. On the other hand, a full-sized apple tree can take 6 to 10 years to bear its first fruits and they are prone to disease.

Fruit trees can attract rodents, bugs, and they require special care. If neglected, the fruit can fall to the ground and rot, attracting flies, fruit rats, and other sorts of pests. It's important to understand that these trees require your attention and if you do take care of them, they can produce an abundant crop of fruit. If you don't, they can be a haven for pests and create a rotten mess in your yard.

**You should not compost citrus fruits. It can make your compost too acidic.

You can eat fresh fruit, dehydrate it for long term storage, and use it to create things such as pie-fillings, jams, apple butter, applesauce, cider, and more. Many types of fruit provide several different options for their use, with apples being one of the most popular. If your family doesn't love apples, you can also sell them. Your backyard homestead could easily supply enough to operate a seasonal produce stand in your area.

Apples, peaches, nuts, cherries, lemons, oranges, limes, figs, and pear are all very popular trees for those hoping to grow food. They all have specific zones that they do well in and some that are available in many different varieties. In the south,

For more unusual additions, consider the pawpaw tree or the guava tree. Pawpaws are grown in southern regions and very popular. They have been described as having a taste similar to a banana. They grow well without a great deal of special care.

Testing Soil

You don't need to spend money to test your soil. The vast majority of soils in the United States are low and only a few very rare regions are high in pH. The best way to raise the pH of the soil is to add lime. Lime is natural and is good to add to your soil. You should not add lime at the same time as a chemical fertilizer, but we've told you a myriad of ways to avoid chemical fertilizers and weed killing already.

When you can avoid chemicals, you should. Lime will raise the pH in your soil if it is very low. Composting and layering compost onto the places you want to grow vegetables in is typically enough. If, however, you wish to check your pH levels, you can. There are kits that you can use. You can also send soil samples to your local Department of Agriculture in some states.

Most plants thrive in slightly acidic conditions. That will fall somewhere between 5.5 pH and 7 pH. There are ways to test your soil without a kit and without spending money. Here's one method:

Place two teaspoons of soil from the area of your garden area into two separate cups. Add a half cup of clear white vinegar to the first cup. Wait to see the reaction. If it fizzes, your soil is alkaline and above 7 pH. You don't need the second cup and your treatment is lime.

If it doesn't fizz, you will move to the next cup. Add just enough distilled water into the second cup to make your two tablespoons of soil turn into a muddy mixture. Now, add 1/2 cup of baking soda. If it fizzes, your soil is acidic. It's likely 5-6 pH. Your fix is adding bone meal to lower your acid. It's a slower process because you must add it slowly. It can take a season to get your pH back to the right balance. It depends on how far off you are.

Obviously, if you got no reaction in either cup, your soil is in good condition, a balanced pH that should be wonderful for most plants. You should do a little homework and learn what plants do best in high or low acid soils.

You'll have a great idea as to what will thrive in other parts of your yard as well. It's also important to understand that not all areas in your yard will be the same pH all of the time. It may pay to test more areas to see if there is a better garden location for you.

The macronutrients that are most important to plants are nitrogen, potassium, and phosphorus. Calcium, magnesium sulfur and carbon are also very important to a healthy growing environment. Then water and our air both supply hydrogen and oxygen which are, of course, necessary.

Avoid walking on your garden beds so they don't get packed down too hard. The soil needs to be aerated so that the roots are getting water, nutrition, and oxygen. When soil gets packed and then wet, the water can force the oxygen out and away from the plants. You want to avoid this by keeping your soil loose.

If you see yellowing between the veins in the leaves of your plants, you've likely got a magnesium deficiency and you can add Epsom salt to correct this problem with your soil. This is also known as magnesium sulfate and you want to add about 5 ounces of it to 100 square feet of garden area.

The vast majority of nutrients will be fixed when you apply compost to your gardens, and this is why successful gardeners use it so readily. When you can't make it, buy it. Start learning how to make it and then never look back. As you continue to use compost, random pH checks should show you that your soil is healthy, and your plants should thrive.

Macronutrients: Essential for Our Health

Macronutrients are trace elements that not only create healthy plants, but they also give us the nutrition that we need from our foods. The healthier our garden plants are, the healthier we are when we eat those foods. It is a fact that when we eat a nutritious diet that is well-balanced with fruits and vegetables, we are healthier people.

Calcium and iron are pretty easy to get mixed into your soil through good composting practices but there are some things that you might find are in short supply and have no idea how to add them. Rain and erosion remove them over time and overgrowing can remove many of the necessary trace elements over time. How do you put them back?

You can try getting kelp to add to your compost. This is a good source of things like fluorine and strontium but another source, which is very simple, is to add ocean minerals that are rich in all of the necessary nutrients and elements that will help give your food the most nutrition possible. Seawater contains every known, naturally occurring element on the planet.

Sea minerals can be purchased for adding to your garden. They come in many brand names, everything from Miracle-Gro to Sea Magic Plant Stimulant. They're even very reasonably priced. You could surely use ocean water you think, but no. Don't do it. Salt is the bane of gardens and soil. In fact, salt in the soil can ruin your soil for generations. It is the equivalent of sugar in a gas tank. It will ruin your ground.

The minerals in the water, aside from mass quantities of salt are wonderful, though. Buying sea minerals will solve the vast majority of nutrients lacking in your soil. It takes very little when mixed in water to spread on your garden as they only need to be added once per year and a five-pound bag could last you for a few years with a small garden.

There is one nutrient that is not included in sea mineral additives for your garden. This would be boron. If you have hollowed out stems in your broccoli or potatoes with gray centers, you're probably in need of boron in your garden. This is a very simple fix and you should only add a small amount.

One teaspoon of Borax in a gallon of water should be spread over an area 4 ft by 8 ft. Then water this area as you normally would with just water. You only need to do this Borax feeding *once per year*, at the start of your growing season.

Starting Seeds and Creating Strong Seedlings

The way that you start your seeds is important because they must sprout and open successfully. Seedlings, the plant that has sprouted, must be kept inside and protected until they are ready to 'harden off' before being transplanted outdoors. Giving your seedlings, this head start will make them stronger plants once in the garden. It's like pre-k for the plants before they graduate to the garden.

Ways to start seeds:

A necessary ingredient for growing seeds is sunlight. You'll need to have your seeds in the sunshine by placing them in a windowsill or by using an indoor UVB light.

Water - Some seeds can be started with just water. An example of this is an avocado seed that is placed over the top of a cup and held in place with toothpicks with only the bottom of the seed in the water. A root will begin to form in a week to two weeks.

Paper towel moistened in a Ziplock bag - This method works really well as the Ziplock bag works like a greenhouse and the moist towel will give the necessary water to keep the plant seed supplied with water. Once the seed sprouts, you may transfer to a container filled with soil using a method below and then continue growing your seedling indoors.

Small eggshell or cups - When you start seeds in half an eggshell or in a paper cup, it can be placed into the soil when the seedling is ready to transplant and the roots will grow through them, and the organic containers will compost in place and feed the plant.

Planted in 1-inch pots - This is a method most traditional in which seeds are placed in a growing medium and watered, kept indoors, and in a humid environment. Once the seeds sprout, you'll follow the normal growth and hardening off techniques detailed below. Instead of plastic pots, you can use paper egg containers that you may tear off and place directly into the soil. Some people don't like this method, saying it wicks too much moisture from the seeds and seedlings.

The first year that you plant a garden, you might try a few plants started in each way. This will help you learn and decide what method you appreciate most and least. Sometimes, we just don't like a method because it is awkward for us. Other times we fall in love with a method that feels fool-proof to us. Go with what you love. Remember, it is *your garden*. There are as many methods of gardening as there are people.

Once your seedlings reach a height of a few inches, you should begin to strengthen them by placing a fan near them and have it blow on low. Turn your plants as they begin to grow slanted away from the fan or toward the sunshine. This will help the main stalk of each plant to grow strong and be ready to handle the outdoor breezes when transplanted. Think of this as training for the real thing.

Tomato seedlings are special. They will begin to grow new roots sometimes, just above the soil. Watch for these. When you see them, like tiny nubs, add more soil to cover the main stalk up higher. This will help them root deeper and be stronger.

Seedlings shouldn't be given fertilizer for the first two weeks of their lives but keep them moist. A spray bottle is a wonderful way of keeping them just wet enough without overwatering. Try to avoid using paper cups that are thick and absorb the water, wicking it away from the plants. These are the type from egg cartons. The premise for them is good but they can dry too quickly because they wick the water away from the seed and the soil. If you know you won't water them 4 or 5 times per day with a mister, don't use them.

In fact, starting seeds in cups that are placed directly into about a half-inch of water is far superior. This allows them to draw water up to them when they need it. This will help them most and be far more convenient for you as well. Adding a bit of water once or twice per day should be plenty.

It's important to note that not all things should be started as seedlings. There are some things that do best when sown directly into the soil where they will be grown. These are typically root plants such as beets, carrots, turnips, corn, and grains. When you have any doubts as to how you should start a plant, the internet is there with answers. It's also a fantastic idea to join social media gardening groups.

It cannot be understated how much information you can gather from people who are in these groups with years and years of experience that simply enjoy sharing their knowledge.

When you have a question about specific bugs, plant companions, determining the source of your bug damage, and more, these groups will offer you a plethora of answers as long as you understand that not all advice should be followed. You'll have to learn to sort the good from the bad.

Community gardens are also a wonderful learning experience that can't be overstated. You can work alongside others, share in the work, and then get a portion of the garden growth each season. If you're looking for a way to learn without commitment, this is a great way. You don't have to completely dive into tilling your entire backyard to start either. Pick a corner, grow your garden a little each season. Too many people get in way over their heads by taking on too much too fast. Go easy on yourself, this is a learning experience.

When Should Seeds Be Started?

Indoor seedlings should be started between 2 and 12 weeks prior to the last frost. If you expect to plant in April, start seeds as early as the end of February and you'll be prepared. Obviously, someone in Minnesota isn't going to start their garden at the same time as someone in Alabama. Know your zone and plan accordingly.

To know when to expect your last frost, you'll need to know what zone you live in and then refer to that zone information for the best planting dates. Then you'll know you should start your seeds as much as 12 weeks before that. Many farmers have relied on the Farmer's Almanac for generations. Consider looking at the latest copy.

Those who live in colder climates may not be able to put their gardens in until June of each year, but that is why you should look at the section Extending Growing Seasons. This will teach you how to get your plants started even sooner and how to keep them growing for weeks after the first frost of the fall.

In some parts of the country, a greenhouse can help you grow year-round. It's important to know all of your options and get your seeds started as soon as you can. The sooner you start each year, the better. Get a jump on it. The early bird gets the worm!

Why do you want to start seeds as soon as possible each year? For one thing, you can beat the heat. There are many plants that don't fruit when the summer heat is blazing. Tomatoes and peppers don't like it when the heat is consistently above 90-degrees, for example. The plant may look healthy and green but stop producing fruits.

Until the temperatures are reduced, they will not fruit again. For this reason, starting early will give you one harvest, and then when temps reduce in late August to mid-September, you'll be able to gather a second crop before freezing. This can be the difference between 100 pounds of tomatoes or 300 pounds of tomatoes in any given season. That's a lot of canned tomatoes, which are a staple crop due to the many ways they can be used.

Tomato Products:
- Stewed tomatoes
- Spaghetti sauces
- Marinara and pizza sauce
- Soups
- Pickled tomatoes
- Tomato and okra
- Fire-roasted tomatoes.
- Soup base
- Salad tomatoes

These are just a few of the more common ways that tomatoes can be processed and preserved. Like potatoes, they are often a garden staple because there are so many different things that can be made from them. Think of the foods that you will use the most and focus on those first.

Seedling Start Times Before Planting Outside

Artichoke	8 weeks
Basil	6 weeks
Broccoli	4 to 6 weeks
Cabbage	4 to 6 weeks
Cauliflower	4 to 6 weeks
Celery	10 to 12 weeks
Collards	4 to 6 weeks
Corn	4 to 6 weeks
Cucumber	3 to 4 weeks
Eggplant	8 to 10 weeks
Kale	4 to 6 weeks
Leeks	8 to 10 weeks
Lettuce	4 to 5 weeks
Melons	3 to 4 weeks
Okra	4 to 6 weeks
Onions	8 to 10 weeks
Peas	3 to 4 weeks
Peppers	8 weeks
Pumpkins	3 to 4 weeks
Spinach	4 to 6 weeks
Tomatoes	6 to 8 weeks
Watermelon	3 to 4 weeks

This list is to give you a general idea. Most packets of seeds will have the time to plant on the back of the packet or you can look the information up for specific plants that you are interested in growing.

As you can see, planning a garden must be done months in advance to ensure your success. Most successful gardeners begin planting the garden for the following year as soon as they harvest this year's garden.

How to Water Your Seeds

Seeds can easily be overwatered, underwatered, and dislodged from the soil they are in by pouring water over them. The best way to avoid all of these situations is to place seedlings into a tray of water with enough water to cover up to the bottom of their containers. At the end of the first two weeks, add some liquid fertilizer to that water and add water as necessary to the water tray each day.

The tiny potting methods used for seedlings will dry out very quickly. Poke holes in the bottom of their containers if you've made them yourself. Allow them to sit in water so that the seedlings can draw water up from the bottom without being disturbed or drying out. A seed that has opened and then is allowed to get dried out will die.

Pulling water from below is the natural way for plants to grow and will encourage root growth, helping to make your new starts strong. One thing to watch is the appearance of mold on top of the soil. There is more than one type of mold. If you see a gray fuzz, move your plants to direct sunlight in a southerly window for two to three hours per day. Cut the water back a bit as well. This will generally solve that issue.

Plastic domes sold as 'portable greenhouses' can also cause mold to grow because of the atmosphere that they encourage. Take the lid off and allow your plants to breathe. This is generally all you need to do to solve the issue.

Hardening Off Young Seedlings Before Transplanting

This process is meant to strengthen your plants to be able to survive outdoors. It prepares them gradually for living outside in their new environment.

Approximately two weeks prior to transplanting your new seedlings outdoors, begin taking them outside for just a couple of hours each day for the first two days and gradually build the time each day over a two week period until you can leave them outside for a night or two just prior to transplanting them to the garden.

This acclimation period gives them time to grow a thickened stalk that can resist a breeze and to adjust to night temperature changes and full days of sunshine, surrounded by other plants.

When you transplant them, ideally you have used containers that will compost when placed directly into the ground. This allows you to plant them without any disruption to the roots at all. If you do need to transfer them from the container to the ground, give the container a gentle squeeze as you turn it upside down into your hand. The roots will slide out with the soil in one piece most generally.

Dig a hole large enough to accommodate this entire root ball and soil and place it gently into the ground, pulling some soil up and over the top, slightly higher up the center stalk to give it support and promote the growth of new roots that will secure it into this new soil. Be gentle and water it liberally after you've placed it.

Continue to water them each day well before the rays of the sun are directly upon them and be careful to not water the leaves, only the soil at their base. Add ground cover at the base of your new plants, such as hay or grass clippings. By adding this type of organic matter, you'll help your seedlings to have water readily available throughout the day, rather than the soil leaching it away.

Continue to fertilize your garden once every two weeks with a good organic fertilizer to ensure that they grow strong and continue to grow. As your plants reach the point of fruiting, they'll need all the nutrition that you've given them in order to develop large, healthy produce that will eventually fill your table and your cupboards.

Soil Preparation and Weed Control Methods

Now that seeds are started, you want to ensure that one challenge of keeping a healthy garden is keeping a garden that is free of weeds and creating soil that is healthy and full of nutrition so that plants will grow to their fullest potential. Not having enough nutrition can cause fruiting plants to be unable to produce fruits.

Before planting directly into the ground soil, you should spend some time conditioning your soil and it is a year-round process to be thinking about. This is because composting and fertilizing are often done through the year so that soil is ready in the spring.

If you live in a place that the soil is particularly bad and you're unsure that using your soil is in your best interest, you may want to look into gardening techniques that can help create great soil in just about any conditions.

No-Till Garden Techniques

This includes:

- **Back to Eden -** This is a no weeding necessary, no tilling gardening method that is said to be an excellent way for beginners to start a garden. For many of you, this will be the most appealing choice for many reasons.

  A very big plus about this method is that it doesn't require constant watering which is limiting for many people who have to pay for their water usage or don't have running water because they live off the grid and rely on the collection of water.

  Keeping a garden for them has been prohibitive because what it saves them in food it has cost them in water use. You want to avoid a water bill or not use the precious water that you're able to collect from the sky. We'll discuss ways of doing this later in the material, but the Back to Eden method will lower your need for watering plants to essentially zero.

1) Determine where you want your garden. It can be on the grass, on rocks, anywhere that is a good location for you because you're not going to be tilling.

2) Cover the entire area with newspapers. Don't leave any gaps for weeds or grass to grow through and put down at least 3 - 4 layers of newspapers.

3) On top of your newspapers, spread at least 3 or 4 inches of compost. You can make your own compost, which we discuss in the section *Composting in Your Backyard Garden*. If you're starting out and have no compost ready yet, you can buy it. Look for local farms that have compost. Mushroom compost is great and when you find a good source, you can often purchase compost very cheaply by the truckload.

4) On top of the compost, you'll layer 6 inches of wood chips. Don't use lumber because that can contain treated lumber and you don't want various chemicals leaching into your plants. You want to use the mulch from trees that will include the leaves, bark, and wood pulp. The smaller the wood chips are, the better.

*If you are baffled as to where to find this sort of product, call your local dump and find out if they have a refuse area for yard scraps that are often picked up separately from normal trash. They will use a woodchipper to grind this yard refuse up and it's there for the taking if you want it, typically at a very low fee for large amounts. 2 to 3 tons is enough for a relatively large garden area. You may find it for less than $10 per ton, depending on where you are located.

5) Plant! You can wait a season and allow the area to turn into a very rich soil, but it is completely fine to place your plants in compostable paper cups right into the top layer of wood chips. It all begins to compost quickly and will stay damp after rain to provide plants with water long in-between rains. Once you have conditioned the area in this way.

6) To keep your beds in good shape from one season to the next, cover them to prevent winds carrying grass and weed seeds into them after you've harvested. Place a couple of inches of fresh mulch in the form of wood chips on top of them each year or add the fresh compost that you've now had a full season to create. The idea is to continue keeping the area filled with organic matter that holds onto moisture so that you don't need to water it. Nature will handle it all for you. Almost sounds too easy but there are rave reviews of this method.

- **Hügelkultur** - It's pronounced HOO-gull-culture. It translates to "hill culture" and it's a German method that relies on building hills by burying organic material and placing soil over the top. It's a wonderful way to grow gardens and is less difficult than it sounds.

1) To begin, logs are placed in the ground in approximately one-foot deep trenches. Start by removing the grass and dig the trench. Save the grass for the compost later. Stack your logs as high as you'd like to create your hills. Old logs laying around your property that have fallen during storms are perfect. The trees are your base. You can also use grass clippings and refuse from your yard to build your Hugel mounds.

2) The logs break down slowly, taking years in most cases. The harder woods take the longest to break down completely. During this process, they are continuously releasing nutrients into the soil, creating a thriving environment for growing nearly anything. The nutrients in a Hugel garden could potentially last as long as *twenty years*!

3) Since this is essentially a compost pile, it will behave like a compost pile and create heat inside as the organic matter composts. That means that you can effectively create a separate microclimate that can extend your growing season for several weeks each year. That means that if you are in a zone with a very short growing season, this might just be the perfect method for you. You'll get your plants started soon and keep them happy for longer each season.

4) An additional benefit of the Hugel garden is a simple concept to grasp when explained. In the first year, you'll want to water the garden well. The logs underneath will absorb and hold water. In fact, they'll hold it so well that you won't have to water your garden in the following seasons more than *once per year*. This makes them an incredibly good choice for people who live in areas with low rainfall and without reliable water resources.

5) The materials to build your Hugel mound are easy to come by in the majority of situations.

- Branches
- Grass clippings
- Newspaper
- Cardboard

- Straw/hay (be careful of those with grass seed in them)
- Leaves
- Compost
- Manure
- Of course...the initial logs

Avoid:

- Black walnut hulls and trees
- Diseased plants
- Old-growth redwood
- Live wood - you must use deadwood to avoid the tree taking root.

6) Add the initial soil that you dug out to cover the mound, in addition to more topsoil to fully cover your mound. Water it well in the first year and begin planting right away.

- **Lasagna Bed Gardening -** This is one of the easiest, cheapest and fastest ways to create great gardens. It's a matter of layering ingredients that will compost in place, providing nutrients to the soil and the plants that will ultimately call your garden home. It's not difficult, nor is it something that requires extensive time. You can put together a lasagna garden in one afternoon.

1) Determine the area for your garden. You don't have to have raised bedsides, but it will help to keep your layers in place if you do have something to serve as a border. Rocks, landscaping timbers, even old logs from your property will work. This will prevent your layers from being washed away in rain.

2) You simply alternate layers of brown compostable materials. Brown layers can be pine needles, leaves, shredded newspaper or cardboard, compost, manure, peat moss, etc. Starting with a layer of newspaper or cardboard will help to smother grass and keep weeds from taking over your garden. Build your layers on top of that and you don't need to be precise with them, you don't need to be concerned with how much of everything you layer, just layer what you've got and try to create a minimum of two feet tall in total. This will shrink.

3) The next phase is called the 'cooking' of the garden. Basically, it's the composting that takes place and creates heat. Your two feet of compost will reduce to just a few good inches in a few weeks. Moisture is an essential part of the process but unless you are in a severe drought, nature should handle the little moisture that is necessary. If you water the lasagna, you'll possibly create rot, which is not the same as compost.

4) To plant your veggies, you may have to poke a hole into the cardboard underneath your layers. That's not a problem and will likely be very soft and easy to make a hole in. Plant your garden as you normally would. You will only need to maintain your garden for the rest of the season. This includes adding mulch such as grass clippings or straw to the base of your plants to keep moisture near their roots. Water when necessary and repeat your layering the next year. In time, you'll have several inches of rich, black topsoil to plant your garden in.

Saving and Storing Your Harvest

There are many ways in which you can prepare your garden harvest to be saved and used throughout the rest of the year until your next garden. Being able to save your foods and preserve them properly is enormously important when you are trying to be as self-reliant as possible.

You can be prepared for the next pandemic, natural disaster, or whatever chaos is thrown your way; you'll always have food security and that is the main issue facing families when chaos strikes. Just imagine what happens if you lose your job, your spouse is diagnosed with cancer, and you suddenly find yourselves trying to juggle all the bills and eat. How do you feed your kids?

Knowing that you have a pantry that is filled with a year's worth of canned goods and dry staples will take a thousand pounds from your shoulders at a time when you need to focus on other troubles. Many families are what we refer to as food insecure.

Missing one paycheck can send them to the local food pantry. If you've ever been in this position, you'll understand the feelings of helplessness. It's embarrassing as well. You will never have to go ask for food ever again when you follow the principles of this book.

There are a few different methods for storing your harvest and there are some that are complicated while others are quite simple. Choose what works best for you in your situation.

- **Canning** - This is one of the oldest methods of preparing foods for the dry good shelf in your pantry. Canning can be done in a few different ways, some of which are specifically called for, given what you are canning. If you are going to try pressure canning, a word of caution: Do not do this until you've taken a class. Have a mentor to work with you.

  Pressure canning is dangerous. It is possible to blow up your kitchen. A story comes to mind of a woman who was traumatized because her pressure cooker didn't seal properly. It exploded, left her kitchen destroyed, blew cabinets off the wall, and killed her dog that was sleeping in the kitchen. It is not to be done by someone without experience.

The process, in general, is to use glass jars that come with a sealing ring with rubber and a lid that fits inside of that. Your jars, lids, and rings must all be sterilized in boiling water prior to use to ensure that you are not introducing any germs into your canned foods. This would allow harmful bacteria to grow inside the jar and could potentially poison you.

Common foods that are canned include cooked green beans, potatoes, carrots, stewed tomatoes, diced tomatoes, and sauces made with any of the above. Homemade soups can be canned, homegrown beans can be cooked and canned also.

Canning must be done properly, but once you learn and invest in the start-up supplies that you'll need, you'll be able to store the types of foods that you've been buying in your canned goods aisle of the supermarket. Anything that you buy canned at the store can be canned at home.

- **Dehydration** - Purchasing a dehydrator is highly recommended. You'll be able to dry fruits and vegetables for use later or enjoy as dehydrated snacks that your family can eat at any time. Dehydrate grapes and make your own raisins, for example.

Dehydrating fruits and bagging them will allow you to store them at room temp for a short while, freeze for longer, or even can them for up to a year in most cases. All fruits can be dried. Meats can be dried too. Most veggies can be dehydrated and later soaked in water to be reconstituted for soups and recipes. A dehydrator will get a lot of use.

- **Freezing** - Many vegetables are wonderful when frozen and will remain good in your freezer for a year or longer. Take care to protect them from freezer burn. Most vegetables should be blanched in boiling water for up to 10 minutes but as little as 30 seconds for some. It depends on the food you are preparing to freeze. Here are a few examples:

  - Asparagus - leave whole or trim - 2-4 mins
  - Green beans - trimmed or snapped - 2-4 mins
  - Bell peppers - cut into strips - 2 mins
  - Broccoli - cut into 1 ½ inch pcs - 3 mins
  - Brussel sprouts - whole or halved - 3-5 mins
  - Cabbage - wedged or shredded - 1 ½ mins
  - Carrots - ¼ inch slices - 2 mins
  - Cauliflower - 1-inch florets - 3 mins
  - Celery - 1-inch lengths - 3 mins
  - Collards - whole or chopped - 3 minutes
  - Corn - whole ear - 7-10 mins (Can cut from the ear after)
  - Okra - trim ends - 3-4 mins
  - Kale - stemmed - 2 mins
  - Summer squash - ½ inch sliced - 3 mins
  - Tomatoes - whole (peel after blanch) - 30 seconds
  - Turnips - ½ cube - 2 mins
  - Zucchini - ½ inch slices - 3 mins

- **Pickling** - This is typically done prior to canning and is a way you can provide some variety to the things you grow. Using cucumbers, beets, and even eggs gathered from your chickens and hard-boiled, can be pickled and canned. Don't forget that you can make loads of sauerkraut from cabbage and it's very easy to do with salt and vinegar.

- **Fermenting** - This includes making sauerkraut and kimchi. Fermentation is the method by which carbohydrates are mixed with sugar or starch to go through a metabolic process (fermentation) to create alcohol or acid. Yeast creates this process in the making of bread dough, which causes the bread to rise. Fermentation has been used for centuries to create wine by fermenting grapes with sugar. It's a wonderful way to preserve foods as well.

There are more methods of saving foods that involve crushing and creating your own ground spices as well. You can and should also learn to save some seeds to grow your next season's garden. This is an additional way to save on costs and ensure that you get the same quality each year. When you aren't happy with a plant, don't save those seeds and try something new until you are happy. In time, you'll have more than enough harvested seeds to share with others and perhaps even sell if you like?

Consider joining some social media groups that are centered on gardening, canning, and seed sharing. You'll find that there is a wonderful community of people out there who are willing to help guide you and point you in the right direction. Seed sharing is a wonderful way to share your seeds and get others in return so that you can add to your garden each year as well.

Extending Growing Seasons

One of the biggest things that you can't control when growing your own food is the weather. Some parts of the country have shorter growing seasons, thanks to their cooler climate. There are ways, however, to extend growing seasons and when you use them, you can grow hundreds, if not thousands, more pounds of food each year.

Consider a Winter Crop

If you have livestock, even small things like rabbits, growing some things that you can plant as a winter crop. Wheat, barley, and oats are winter crops. They are sown just before winter and harvested in the early spring. Some people plant a winter crop merely to till it under in the early spring, to return nutrients to the soil.

Chickpeas and fava beans are also winter crops. Consider planting something that you can use as feed for animals or for your own food stores. This will add to your growing ability by adding an entire season of growth to your crops. Planting crops that can stay in the ground and sprout in spring is an excellent use of time, getting the most for your money.

Use Cold Frames

Cold frames are essentially like miniature greenhouses that can be formed around plants to keep them warmer and extend the growing season. Many ways of doing this include using straw bales around the plants and laying an old window over the top to act as a magnification of the sun. The straw will protect them from cold wind and be excellent insulation.

Cold frames can also be constructed, using wood and rigid insulation board inside the walls and using a window or piece of plexiglass as a lid that can be opened and closed by adding a hinge.

Build a Hoop House

These are essentially a portable greenhouse and can be fashioned easily with PVC pipe and clear plastic sheeting. The PVC can be placed into the ground on either side by creating a hole with a larger piece of PVC or metal pipe.

Slip the PVC into the hole on one side and gently bend it into an arch on the other side until it slips into the opposite hole. Creating several of these hoops and using some lumber to frame them on each end, into a structure that will withstand some winds, adding a door, and then covering them with plastic will finish them off.

A hoop house can be built around existing crops or used as a greenhouse to have potted plants inside. A well-built greenhouse or hoop house with insulation can remain as much as 30-degrees warmer than outside air. This can easily extend your growing season by several months.

Utilize Compost Creatively

Compost naturally creates heat from the chemical processes that are happening to break down the organic materials. Compost piles can reach internal temperatures of 120 to 170-degrees Fahrenheit. Using this by covering a compost pile inside of your hoop house can increase the inside temperature to utilize it even in the winter.

Compost piles kept underneath an elevated rainwater barrel will prevent it from freezing so that you can have that to water greenhouse plants all winter long in many climates.

Covering outdoor water lines with compost and straw can prevent hoses and water lines from freezing so that you can run lines to plants and trees. Compost and straw added to the base of trees will prevent young trees from early frosts that could harm them.

Simply remember that compost = heat. Free heat.

The fact is that you can never have too much compost and you can find yourself addicted to composting everything you can think of, even scavenging grass clippings from all of your neighbors and asking lawn crews to bring you bagged refuse instead of paying to dump it at the landfill. Yes, they have to pay to dump it so why not bring it to you?

Some Plants Do Well with Specialized Growing

One of these that comes to mind is the potato. Potatoes are a staple crop because this one vegetable can be used in so many ways and is also able to be preserved in many ways.

From a few pounds of potatoes, you can make:

- Mashed potatoes
- Baked Potatoes
- Scalloped potatoes

- Hash Browns
- Diced Home Fries
- Sliced American Fries
- Potato Chips
- Dehydrated potato flakes for instant potatoes later
- Dehydrated potatoes for soups later
- Boiled potatoes
- Twice Baked Potatoes
- And so much more…

Potatoes, for good reason, are an important crop to many people. Potatoes are traditionally grown in mounds; they have an extensive root system that develops the potato while the green leafy part of the plant is above ground.

In recent years, it's become very popular to grow this root vegetable in containers, specifically in five-gallon buckets. There is a method to doing this successfully.

One of the most successful ways to grow a ton of potatoes from one bucket is to cut the bottom completely out of the bucket and place it over some newspaper, layered four to five layers thick to prevent weeds from growing up into your bucket.

Add just a few inches of potting soil into your bucket and place your first two or three potato starts. This will be a piece of potato that already has sprouted from the eyes. You can do this yourself from any organic potatoes from your grocery store.

Eat some and determine if you like them enough to grow them. Then, place as many as you'd like to grow into a brown paper bag and stick them in a closet for a month or two before the growing season. When you pull the bag out and open it up, they should have sent runners from the eyes of the potatoes. Simply cut them so that you've got one eye per seedling.

Water your potatoes and watch them each day. As soon as they sprout above the surface of the container, you can add another two inches of soil and place another two potato starts. Continue doing this until your bucket is full.

Then wait. Continue to care for them and don't let them dry out but make sure that they are also not so wet as to rot the potatoes. This is the number one reason that potatoes don't do well, most people don't allow them enough drainage -- but you did when you cut the bottom from your bucket.

It takes 70 to 100 days for most potatoes to grow to maturity. This is where your bucket method is so very different. You'll harvest your potatoes from the bottom of the bucket by slowly and carefully sliding up until the potatoes are exposed at the bottom of the bucket. If they don't look large enough, carefully slip the bucket back down and wait another two weeks. Wait a week between each slide upwards to harvest potatoes until you've reached the last layer.

Planting several buckets in this way can yield hundreds of pounds of potatoes each season. Saving them can be done by canning them, dehydrating them, and storing them in a homemade root cellar.

Quick and Portable Root Cellar for Root Veggies

You can create storage for root vegetables that will keep them over the winter months. This will work for potatoes, onions, garlic, carrots, and other root vegetables. You'll want to store onions and garlic separately from other vegetables as they will absorb the onion smell and taste.

Purchase metal garbage cans. You cannot use plastic as rodents can chew into them. You'll want the galvanized metal garbage cans with lids. Dig a hole in your yard that will allow you to drop the can all the way down into the ground, allowing just the rim and lid to be above the ground.

Drop your veggies down into the can where they will be kept cool and dry all winter, below the frost line. Place the lid on the can, lay a sheet of plastic or a tarp over that, and lastly place something very heavy over it to keep animals from removing it. Rocks, cement blocks, or anything you've got laying around the yard that is heavy enough to keep them from lifting the lid will work great.

Place a post with a flag on it if you get snowfall. That way you can keep track of where to find your cans when you need to bring in some potatoes or other produce. You might want to shovel a pathway to your cans and keep them close to the house so you can get to them easily if you live in a place that has heavy winter snowfalls.

Saving Seeds and Why You Should Do It

The main reasons that you should save your seeds are that:
- You know the quality of the produce
- You know these plants were grown organically
- They've already proven to be up to your standards
- It will save you money on buying seeds in the future
- You'll always have plenty of seeds when you need them.

Seeds can be gathered from your vegetables, some more easily than others. Plants that are easy to gather seeds from include tomatoes, beans, peas, and most peppers. The seeds don't require any special treatment or handling. Simply gather them and save them. Once dry, place them in glass jars or paper envelopes that you can label with the type and the date. Seeds don't stay viable forever so dating them is wise. Try to use seeds before they are three years old. You'll have better yields.

Biennial crops mean that they must be grown for two seasons to produce seeds. Crops such as these include beets and carrots. They must be left to grow a second season, and, in that season, they will produce all of the seeds you could possibly desire. Still, the weather in some parts of the country won't allow plants to live through two seasons, which makes it more difficult.

Open-pollinated plants will reproduce easier from gathered seeds. Heirloom seeds are often open-pollinated. Hybrid seeds don't tend to do as well so they aren't typically worth gathering seeds from.

This is one reason that so many avid gardeners choose heirloom varieties. When they find what they like, they can continue producing these plants and even cultivate them for seed sales.

Choose the plants you wish to start with based on your long-term goals for gardening and seed gathering.

Many plants, such as corn and vine plants have both male and female of the species and these are far more difficult to keep seeds from that will be good producers. It's virtually impossible to keep these seed strains pure and most gardeners simply choose new corn seed and new cucumbers each season for this reason.

When you choose bean seeds or other types of seeds to keep and grow the following year, be selective. Only choose the seeds from the best plants and the nicest vegetable produced. This will help ensure that the best genetics are passed on in the next generation of open pollinators.

Crops Cultivated Outside of Gardens

Taking advantage of your surroundings is a wonderful way to add to your ability to put food on the table and/or raise money from your backyard homestead. Here are some ideas for things you can encourage to grow in or near your backyard.

Mushrooms

Mushrooms can be a big cash crop if you have success. They don't need a ton of space, depending on the type that you cultivate. Mushrooms are grown from spores, which can be purchased. Some mushrooms grow on old logs, which can be stacked. Others will grow best in bags that are filled with spores and a growing medium and kept in a dark garden shed. There are many methods.

How to Grow Mushrooms

There are two basic ways that mushrooms are grown and neither way is overly difficult but the easiest method, especially if you are new to gardening is the log method. Why would you want to? Mushrooms are a decent cash crop and they are also an addition to your table.

Growing mushrooms in logs is a process done by ordering plugs that contain the mushroom spores. You drill holes in logs. Choosing the right wood is important. Shiitake mushrooms are especially fond of sugar maple, but most hardwoods will suffice.

Adding several holes in logs, adding your plugs to them, and then stacking the logs in columns, so that there are spaces between them, is one of the simplest ways to start your mushrooms. You need to keep your logs moist. You want to start with healthy wood that isn't rotten or decaying. Keep them in a shady, warm, humid area and that is where you'll stack them after inoculation. Water them every two days.

It will take as long as 1 to 2 years for your logs to begin producing mushrooms, but the most beneficial thing is that they will then keep producing mushrooms for as long as 7 to 8 years from the time the first ones appear.

Types of Mushrooms that Can Be Grown in Backyard Homesteads

- **Oyster Mushrooms** - these can be harvested the same fall after you inoculate them. They are one of the fast-growing varieties. There are several different varieties of oyster mushrooms that vary in both color and flavor. They're a favorite of restaurants and sometimes you can sell them to suppliers. Once they begin growing, they can produce fruits from spring all the way to fall each year, for several years. This variety prefers beech wood.

- **Shiitake Mushrooms** - This is the second most popular mushroom on the planet. They're used most frequently in Asian foods. These take up to 2 years to fruit for the first time and are well-known for reducing cholesterol and boosting the immune system.

- **Lion's Mane Mushrooms** - These have a unique look that resembles mane that flows, and they are ivory or white in color. They will grow readily in beech, elm, poplar, or maple wood. These will do best in logs that are large - circumference of ten inches will be best for them. The bigger the logs the bigger the mushrooms will grow. This variety takes 1 to 3 years to begin fruiting, is medicinal but can also be eaten.

- **Reishi Mushrooms** - This is not an edible mushroom and is grown for medicinal purposes. It can be dehydrated and ground down into tea. These take 2 years to fruit but also won't fruit in logs that get too dry, so stacked logs don't work as well for them. In their case, logs should be kept on the ground itself.

- **Chicken of the Woods** - This is a variety that can be eaten. This loves to grow on oak woods, and they grow in clusters that can get very large, up to a foot in diameter. They are very bright orange and yellow in color, look like a shelf growing on the tree, and in the wild will be found halfway up the trunk of a tree. This is a flavor favorite for many mushroom lovers. The problem is that this is one that is harder to grow, and the method is still not perfected. It's more often found in the wild.

Mushrooms are healthy, add to meals, and they can be dehydrated for storage or ground into a powder that can be used in teas and gravy. Some people use them to create additions to coffee or tea. They are rich in nutrients that can be added to your diet but are virtually calorie-free.

Honey Collection - Having a Beehive

This is in the garden section because fruits and veggies need pollination and bees provide that. You'll need to check with your local ordinances to find out if a hive is okay where you live, but in most cases, it will be. Find a local beekeeper's group and they'll know too.

A hive will allow you to keep bees that will help your garden flourish and will also provide you with hours of fun as you keep an eye on your growing backyard family. When they work hard and have plenty of collected pollen, they'll fill the hive with honey which you can extract and have for your own use.

Your garden success will depend on the bees doing their job and they'll depend on you to give them a happy, safe home where they have water and nourishment, protection from the elements, and a hive that is clean and sturdy.

There is a lot to learn about beekeeping; it could certainly be a book in and of itself. Look for beekeeper clubs and organizations near you. You'll likely find that there are many. In fact, many volunteers will come and help you set your hive up and help you with learning how to care for them and collect honey.

There is an initial investment in the pieces of your hive, a *nuc*, which houses the bees, and the bees themselves. You'll want a protective beekeeper's set of clothing that includes a hat and face screen so that you aren't stung. You'll learn how to use smoke to sedate the bees into a calm state that allows you to handle them and extract honey or just to look in on them when you need to.

You'll also need to learn how to properly winterize your hive and keep your bees happy year-round so that they don't swarm and leave your hive. Beekeeping is a rewarding experience that will provide the world with very important pollinators that need our help right now because honeybees are in trouble. They are endangered and yet very necessary.

That means that keeping bees is doing a good thing for the entire planet because, without them, all foods that require their help to pollinate will become impossible to grow. They will provide you with delicious honey in return. Make sure to plant some bee-friendly flowers as well, to completely round-out your backyard habitat.

Basic Steps to Having Your Own Backyard Beehive:

Bees need four things:

1. They need sunshine for warmth and some afternoon shade if you are in a very warm location.

2. You'll need to provide them with a source of water as close to their hive as possible. Bees need plenty of water and you can provide them with a water bottle designed for bees, a birdbath close to them, filled with some marbles in the water so they've got a place to land, or another source.

3. Protection from the wind is also imperative for the hive. Planting evergreens or installing windbreaks near them will be especially helpful.

4. Bees need privacy so they'll need to be in an area of your yard that is relatively quiet and calm the majority of the time. They don't want to be where the kids play, for example.

Bees should be installed into new hives in the spring. This is the best time for them to move into a hive and get right to work, collecting honey and making themselves right at home.

Supplementing them with some nectar until they are flourishing in their new digs is a good idea. They've got a lot of work to do, establishing a new hive, after all. Use feeder lids on jars and fill them with half water and half sugar mixture. You'll be shocked at how much they can drink in a day but after the first three days, it will taper until they aren't needing it after a couple of weeks or so.

You should inspect your hive regularly, but not overly often. You want to ensure that you don't have a pest problem, that the queen is present and that she's healthy, and that your colony is thriving and making honey.

You can, and should, join a group of beekeepers to learn the ropes. Enlist their help in setting your hive up, introducing your new bees to their hive, how to inspect your nucs and ensure that your bees are thriving. There is a lot to learn and the best people to learn from are those who have had bees.

Local colleges sometimes have programs for beekeeping too. There are social media groups that you can join and find local people to network with. One thing that you find is that beekeepers are passionate and enthusiastic about beekeeping. They love to help new people become as addicted to bees as they are.

When you keep bees, you do your garden a favor and you do the world a favor. It's just another step in connecting to your land and the process of self-sufficiency.

There are a few types of hives, including some that you can build on your own. Most successful beekeepers use a hive that is purchased and put together at their locations. A bee nucleus consists of a queen and a few worker bees. Often called a nuc, it can be ordered and shipped to you, where your job is to introduce it to the hive you've taken the time to put together and place in exactly the right location.

You will need to take care of your bees because if you don't, they'll leave and there goes your honey. Honey is an excellent addition to fresh baked bread or biscuits. It's also a natural way to reduce allergies, by taking a teaspoon of honey each day. Honey is also a natural form of antibacterial that can be applied to a cut under a bandage. Many people use honey in making natural soaps and skincare products with goat milk.

The beeswax can be used to make candles and seal jars. The bee is a wonderful addition to your backyard homestead and nothing that you should be afraid of unless you are allergic to bee stings.

Chapter 2
Raising Food

Small-space Livestock to Put Meat on Your Table

If you are a vegetarian, more power to you. You'll be able to get by without this section at all, but most people do eat meat. It also happens to be one of the biggest costs in the food budget. Not only that, but the meat you purchase at a grocery store is also often loaded with steroids to keep animals healthy and grow faster.

Dyes are used to make meat look fresher. Hormones to boost growth are often injected into chickens at commercial hatcheries. Yuck. No one needs that and no one should *want* that. The animals are often mistreated too and that isn't the way things are meant to be. The relationship with animals should be that they are treated well, revered for providing your family with meat.

Raising your own meat doesn't have to take up a lot of space. If you've got a backyard, there's a high likelihood that you can get by with some forms of livestock that you can raise for meat. If you are squeamish about butchering your own animals, you can pay to have a butcher/meat processor take care of that for you.

You just have to pack them up in crates and haul them in. They'll handle the rest and call you when the neatly folded white packages are available for pick-up. If you're going to go this route, raise more than you need so you can sell some of your animals to offset the cost of the butcher.

There are more options in raising your own meat than you had probably considered before. Here are a few options for you to consider.

Chickens

Many towns and cities are zoned to allow chickens now. The one caveat is that you don't have a rooster. They crow and make a lot of noise and even if you live in the country but have a close neighbor, you might want to forego having a rooster to keep peace in the neighborhood and happy neighbors. Plus, some roosters can be downright mean.

The good news is that you don't need a rooster for a few hens to produce eggs. Chickens begin laying eggs at roughly 6 months of age, though some breeds may take up to 8 months. You can start with chicks that will require a heat lamp, a safe place indoors until they are old enough to be in a well-built run with a coop that will keep predators out. Raccoons, squirrels, stray dogs and cats, foxes, coyotes, snakes, and possums will kill young chickens and full-sized hens. Many will also steal eggs.

Provide your hens with laying boxes that are filled with clean straw, keep their hen house cleaned, and remove and keep their poop for fertilizer on your garden (we'll cover that in Composting). Feed the best quality layer's feed that you can purchase. Learn all you can learn about chickens and join some social media groups where you can ask questions and learn more as you go.

The average hen will lay eggs very well until they are 18 to 24 months of age. At that point, they will go into their first molt. When they begin to lose feathers at that age, they'll also stop laying eggs. This is normal. You've got two choices at this first molt: Keep the hens until they are finished molting and begin laying eggs again or cull them and make meat birds for your freezer out of them.

Many people will keep the best layers and allow them to stay on while culling the ones who were not as productive. Either choice is a win. You'll continue having fresh eggs or your freezer will be filled with meat for soup, stew, casseroles, and fried chicken. This is a good place to note that there are meat bird breeds and egger breeds. There are also chickens known as 'dual-purpose' birds. Know what you are getting.

A meat bird will have larger legs and sometimes grows so fat so quickly that the breast feathers wear off from dragging the ground. They are easy to distinguish from other birds once they are growing and getting feathers, but as chicks, you cannot tell. This is one reason that you should be careful from whom you purchase birds.

Be leery of the dirt-cheap chicks. They might turn out to be all roosters. When someone is selling hens that are 6 months old, you very well could be getting 18-month-old hens that are starting to molt and have already stopped laying eggs. Ask around and find out if they've been dependable sellers before. Social media groups are very quick to call out the locals who aren't honest bird sellers.

The ideal chickens for a backyard homestead will be dual-purpose chickens and you don't have to have a rooster for eggs but to raise some of your own chicks for meat later, you do. It is very much in your own interest to start with a young rooster and hand tame that little guy. Teach him to enjoy your company and that you feed him treats and are not a threat.

An angry rooster has large claws on the backs of his foot, called spurs. They can split you or an attacking predator, wide open with them. There is nothing worse than a mean rooster when you are trying to gather eggs. Many of them end up in the stew pot over their naughty behavior.

Best Five Dual Purpose Hens

1. **Buff Orpingtons** - known for being very friendly and decent size for the meat use after laying their eggs and no longer useful as a layer.

2. **Rhode Island Red** - A really good bird that is hearty and easy to care for. They are a decent size so they will be good for the pot when they've finished their egg-laying years.

3. **Black Australorp** - An Australian breed that is a docile and prolific egg layer, giving you close to 250 eggs each year of their laying years. They're known for being quite healthy as well.

4. **Plymouth Rocks** - This one has been around as long as people have occupied the United States. They've been quite a favorite for their looks and their production. They'll lay 280 eggs per year too, which makes them dependable.

5. **Wyandotte** - This is an attractive bird in many color varieties. They lay approximately 200 eggs per year but are a favorite for their sweet temperament. They are well known

for following you around like a puppy and coming when called. If you plan on breeding them, some of the exotic color patterns are very expensive and make a good side income as show birds to hobbyists. The Blue-laced Wyandotte is a striking bird.

Red Blue-Laced Wyandotte Hen

When you bring new hens home to your existing henhouse, you will need a place to quarantine new birds for as long as 30 days to ensure that you don't bring sick birds into your flock and infect all your healthy hens.

Chickens often need antibiotics, mite treatment, and general care that can be more problematic with birds not adequately cared for nutritionally. Always feed a good quality chicken feed and supplement them with some treats, like mealworms - a treat that chickens treasure.

Butchering Your Chickens

Processing chickens yourself is not overly difficult, though you do need to be very careful to not cut into the intestines of the bird or you'll contaminate the meat.

Most people use a cone to drop the bird into, upside down. This prevents them from running around after you cut their head off -- and yes, they can do that. It bruises the meat too. Using a cone is the best way to dispatch them quickly and without pain. Insert them into the cone so they are upside down. Pull their head down and feel for the ligament under the chin that attaches the head to the spine. Make a deep slice into and across the tendon and then pull the head down but don't pull it all the way off.

Leave them upside down over a bucket to allow all the blood to drain completely from the body. They will wiggle and spasm. This is normal but can be traumatizing the first time you see it. This is also why some people can't butcher their own meat, but that is a choice left to you.

Many people have a ritual of thanking the animal in prayer immediately before or after the act. It's the circle of life and a part of farm living, but also can be a part of your backyard homestead as well, even in the suburbs.

After you've completely bled the chicken and it is no longer moving, you'll need to blanch it in hot water to be able to remove the feathers. Many people use a pot of hot water over a fire but there are scalding pots made for this that can be purchased as well.

The scalding pot is the best choice because you want the water to be at a steady temperature of 130-degrees Fahrenheit to 140-degrees Fahrenheit. You'll need to ensure the carcass is totally submerged and allowed to rest in the pot for roughly 2-3 minutes. You do not want the meat to begin cooking so don't leave it longer than 3 minutes.

Once scalded, the feathers should come out much easier. You can pluck them by hand if you've only got a couple of birds to do. If you plan on processing a dozen or more at a time, invest in a plucking machine. It is filled with rubber fingers. It spins the bird inside and despite spinning very fast is actually very gentle in the way it removes the feathers because the rubber fingers are soft and pliable.

You'll pay about $100 for one, some are more expensive. If you've ever hand-plucked a chicken, you will absolutely be happy to spend $100 to never have to do it again. It's messy, feathers end-up everywhere, and it smells bad. The machine will gather all the feathers and push them out a chute at the bottom that can be scraped into the garbage can.

If you must pluck the bird by hand, you'll want a pinning knife. The larger feathers will come out easily but often leave fine pins behind. The pinning knife is designed to run across the skin and remove the rest of these that are left in the carcass. When you purchase a bird at your grocer, it's common to see some of these pin feathers still in the bird.

Once plucked, you will butcher your chicken by first inserting a sharp butcher knife at the base of the neck to sever the spine. Twist the neck and pull it out of the body. Toss that in a bucket of things that can be used to make a broth that you'll can or use to make soup immediately.

Remove the wingtips and the feet of the chicken by cutting those at the first joint at the bottom of the leg. Make a slice into the joint and then twist it. It should come off easily. Toss these in your neckbone bucket to be used later.

The feet can be dehydrated and fed to your dog if you've got one. They make good organic dog treats. Don't give them neck bones though, they could choke on those. If you don't plan on keeping the feet or neck bones, you should toss them.

You should do this where you have running water ready, just in case you accidentally cut into anything. You'll need to immediately rinse it from the carcass. To begin cutting your chicken for packaging it, you'll begin at the tail.

The small nub of the tail contains a pineal gland that must be removed, and you want to try not to cut into it. Cut up and around it to completely remove the entire tail, gland and all. If you see any yellowish discharge, you've cut into the gland. Rinse immediately and very well, rinse and wash your knife before continuing.

You will want to remove the neck and head first. Remove the head with a good butcher's knife. Simply separate it at the spine, no worries with this part. You can't really do that wrong.

The neck is a bit more delicate as you must remove the bird's crop intact. They have a crop that aids in digestion and there will be bile in the crop. It's a yellowish organ that sits at the base of the neck. Slice into the skin at the back of the neck and work your way down slowly and carefully. Remove the trachea and esophagus from the neck through the slit you've created.

They are the only things in the neck itself. Loosen them with your fingers and feel your way down until you feel the crop. It will be at the base of the esophagus where it enters the body cavity. If you didn't feed your birds for the last couple of days as you should have, the crop will be empty, which is ideal.

Carefully, pull the crop free from the body but allow it to hang, along with the trachea and esophagus for now. Carefully insert your knife about one inch into the bird, just above the vent area (anal area). Carefully cut all along the breastbone and then around the vent on either side. Once this has been separated from the body by your cuts, carefully pull the vent free of the body, and the intestines should remain attached to the vent and come out in one long piece of innards.

The only thing you have left to do is clean the rest of the innards out by pulling them out. There's nothing left to contaminate the meat, so you're fine. Rinse as you go to keep the meat clean and remove the gizzards, heart, and liver to a bowl that can be used in your cooking. You can separate them, or you can freeze them with the bird. Your choice.

You've now got chickens that you can bag. If you buy butcher's bags for this purpose (highly recommended) you can chill your chicken down to a cool enough temperature that you won't overwork your freezer. Then slip it into a bag, dunk the bag back into a pot of hot water for a minute to shrink the bag to an airtight fit, seal it off, and toss it in the freezer. You've just processed about 4 to 5 pounds of quality food for your family.

Chickens also provide you with roughly 3-5 eggs per week, per chicken. Additionally, their poop should be cleaned and put into a compost container for your garden. Chicken poop is a wonderful fertilizer.

Handling Your Eggs

Gathering eggs is one of the primary things that people want chickens for. Many hens will lay 3 to 5 eggs per week, starting at roughly 6 to 8 months of age. They'll do this until 18 months to 2 years of age, at which time they will go through a molt. During the molt, they will stop laying eggs. They will lose feathers and look like a hot mess, as they say in the South.

You do not need a rooster for hens to produce eggs. Many people opt out of roosters for the reason that they are an extra mouth to feed that can also be mean. Some city ordinances do not allow roosters because they are loud, crowing at all hours of the day and wee morning hours.

Reasons to keep a rooster: They provide protection to the hens, will warn you of dangerous interlopers, and they will fertilize eggs so you can raise chicks if you so desire. There is money that can be made from selling fertilized eggs to people who'd like to incubate their own chicks, or from selling your chicks as pullets to those who want laying hens. Young roosters typically end up in stew pots or freezers.

Once their molt is over, they'll start laying eggs again. Sometimes people choose to cull their hens at this time, opting to make stew birds out of them. It's a personal choice but some people will keep chickens well into old age as pets that will continue to produce eggs for as long as five years, though production may slow down some.

Freshly gathered eggs should not be refrigerated. *Do NOT refrigerate your eggs* for maximum life. Fresh eggs will keep at room temperature for at least one full month before you need to worry about refrigeration or cooking.

You can preserve eggs by:

- **Pickle your eggs**. It can be done fast and easily by adding them to pickled veggies, such as beets.
- **Dehydrate scrambled eggs**. You can do this, though it isn't always easy to get them dry enough to store for long periods of time.
- **Freeze scrambled eggs, or raw cracked eggs** in Ziplock bags. You can separate yolks from whites if you like, label the bag with how many egg whites are in each bag for baking later.

- **Prepare eggs** in breakfast burritos and sandwiches, then freeze them.

Rabbits

Raising rabbits is easily done in a small space. They can be kept in a hutch in the backyard and no one will even know you've got them. All you need to start with is a male, known as a buck, and a female, known as a doe. Their babies are called kits.

You'll need to build a hutch. Keep your male in his own cage. Each female should have her own space as well. You'll want to have an additional two spaces to place the newly weaned kits - separating them by sex as rabbits are prolific breeders and will start breeding shortly after being weaned. If you are breeding through the winter, you should provide your rabbits with a form of heat to keep cozy, especially with young kits.

The bottom of your hutch should have hardware cloth that their droppings will easily fall through, with totes or trays underneath to catch that poop to use as fertilizer. Feed rabbits a quality rabbit pellet and do not give them fresh fruits as they can die from sugar shock. Go easy on fresh veggies though occasionally this can be a treat for them.

Kits should be processed into your freezer as soon as possible. Rabbits can and should be butchered very early for the most tender meat. This also saves you from having to feed the extra mouths.

One doe can have anywhere from 1 to 14 kits. A fully dressed, young rabbit will put 2-3 pounds of meat in your freezer as early as

8 weeks of age. You can wait until they are 10 months old if you'd like a bigger broiler.

Most rabbits are butchered between 8 and 12 weeks of age. That could easily put 3 to 30 pounds of meat in your refrigerator. Plus, while you are busy butchering those, the second doe can be well on her way to producing another litter.

You'll never be short on meat. In fact, it's another way in which you can earn a little extra money for your homestead, by selling rabbits or trading for other things. If you've never eaten a rabbit, it's comparable to chicken and makes a delicious addition to any dinner table.

One caveat is that many states have regulations that don't allow you to sell processed meat. You'll have to sell live kits instead if you live in these locations. There are no regulations against bartering your processed rabbits, however. Maybe trade for seeds, or for a new doe from a different bloodline?

Rabbits are easy to process and are prolific breeders so having babies is easy. They give birth in only 30 days. That is an incredibly short gestation period as well. This makes raising rabbits an excellent choice to make your backyard produce the ultimate amount of food.

Butchering Your Rabbit

You don't need to worry about plucking feathers or dunking in cold water. The hardest part of butchering rabbits for most people is dispatching them to begin with. You can opt for any way that works best for you.

Some people hold the rabbit by the back legs and swing it hard against something to stun it and then quickly cut its throat. Some people use a small club to knock them out prior to bleeding them out, while others will use a small caliber pistol and place one shot to the spine, where it meets the head.

As soon as you've dispatched it. Attach the body to nails driving into a tree or on a post you've made beforehand. There should be one nail for each hind leg. Push the foot over the nail, between the tendons of the ankle bone area.

Allow the body to hang upside down, with the head hanging downward. You want their belly facing you. Severe the head at the base of the neck by cutting through the fur across the front of the neck but don't cut all the way through. Carefully cut down to the spine.

Once you've cut down to where you can see those few ligaments that hold the head to the body, You'll cut those When you've finished the cut, let go and the head should fall into your bucket below and allow the body to drain.

As the body drains, cut the front feet off. These can be tossed into a bucket and you can make rabbit's foot keychains or just discard them. Once the feet are gone and the body is drained of blood, use a very sharp skinning knife to cut a circle around each of the hind legs of the rabbit. Once you've done this, cut from each leg toward the center of the body, where you'd imagine the belly button. You will form the top of a "Y" and you want to keep the anal area of the fur intact for now.

At this point, turn the rabbit on the nails so that you are looking at its back. Cut around the lower back until you reach the cuts below the genital area that you made across the belly button area. You've now separated the fur of the upper torso from the genital area. Take hold of the skin on the bottom of the cut and begin pulling it downward. This will take a little muscle until it tears loose. The entire hide should slide off in one piece.

Set the pelt aside where you can salt it and stretch it. Many places will buy rabbit fur from you and this will make you some additional money on the side. Set it aside carefully until you're ready to work with that hide.

To begin removing the meat from your carcass, start with the front legs as the rabbit hangs in front of you. Lift the front legs as you trim through the muscle of the shoulder until you reach the joint itself. A larger knife inserted at the joint should cut right through and give you a nice cut of meat on the bone that looks similar to a chicken leg quarter.

To remove the backstrap, run a sharp knife to just touching the ribs, down each side of the back. You can feel the muscles for the best place to cut. Once you've made these cuts down the side, make a cut across the top and one across the bottom.

Pull down on this piece of meat (one on each side of the back) firmly and you should be able to pull the backstrap from the carcass. You might need to cut a little more here and there until you're very good at this part. You should end up with two separate pieces of meat off the back.

The reason to do the upper body portion of the rabbit is so that if you accidentally puncture the intestines in the next part, it will run down and out, not touching any other meat as everything under it will already be gone.

Turn the carcass around to face you again. Pinch some skin on the lower abdomen, just below the cut you first made to remove the skin. It will seem hollow underneath, but this is where the intestines are, so you want to be careful.

When you lift the skin up, poke a tiny hole, the size of a pea. Insert a finger into this hole and use it to pull the skin up and away from internal organs as you begin to cut downward.

Once your cut is complete, gently push the intestines aside and reach into the body to pull the kidneys, heart, and liver, if you want those. You will probably need to poke your finger through the diaphragm as it will probably be filled with air.

Let the rest of the organs fall into the bucket below or let it get trapped in the ribcage. It won't matter. Your next cut is to remove the back legs. You'll feel for the joint where the leg meets the hip and use your knife to cut the skin until you reach the joint. Pushing the sharp end of the knife through the joint should separate the leg from the body while the foot is still attached to the board nail.

Do the other side in the same way and the body carcass will fall into your bucket. You make final cuts to remove the feet from each leg and you've completed the butchering of your rabbit. Freeze it in bags with as little air as possible. It helps to rinse the meat well first and then soak it in some ice water to reduce the temperature before making your freezer work hard.

Fish in Aquaponics Gardens

This is an option that most people simply don't even think about. It's a fantastic option though if you like to eat fish. Even if you don't, a few phone calls might land you a buyer in a local restaurant or to sell them to the general public if they clean them at home.

Aquaponic gardening is using fish in the same water tank that you use to move water to your plants that you are growing hydroponically, as we covered earlier. The difference is that you add the fish and they will now be what fertilizes your plants, they will provide carbon dioxide to the plant roots, and their excrement is good for the pH levels and the plants.

You do need to feed your fish but only as much as they can consume in a few minutes so that your water stays clean. Some types of fish are best for aquaponics because they thrive in such conditions, grow fast, and are enjoyable for eating. You can use ornamental fish but for our purposes here, only the edible fish are listed.

- **Tilapia** - Grow the fastest. Java, Blue, and Nile varieties do best in Aquaponics. They can weigh 3 pounds in 1 to 2 years. They prefer warm water environments. They are quite honestly hard to kill though.

- **Bluegill** - A small variety that is in the same family as Crappie. They will grow rapidly, reaching 2 to 3 pounds in just over a year. They do best in warmer water but can also be quite cold tolerant.

- **Crappie** - These will reach 4 inches at the end of their first year, 6-8 inches in the second year, and 10+ inches in size if you keep them for three or more years. They do well in warmer water and taste great.

- **Trout** - These species will prefer cold water tanks and produce a fish that has a stronger flavor. Not all people love trout. That said, it's unique and considered a delicacy. They can grow to edible size in as little as 9 months, reaching up to 4 inches in length.

- **Catfish** - These are widely farmed in captivity and are a popular variety that you could sell in excess. Catfish are very hardy and take roughly 1 year to grow to an edible size of 3 pounds. They like warmer water but can thrive anywhere. The Channel Catfish variety tends to grow the fastest.

Of all the species, Tilapia is one of the most highly recommended for their hardiness. It takes approximately 50 gallons of water for every two fish. An above-ground swimming pool can be converted to an aquaponics system that you can float growing vegetation on and fill with as many as 75 small fish, known as *fry*.

75 fish that grow to 3 pounds each is over 225 pounds of fish in about a year. Once your own freezer is full, you can sell the rest relatively easily but try to line a buyer up *before* you commit to that many fish. If your region allows, you can hold your fish up to two years and harvest them as you want them.

Don't feed them live food, stick to flakes and fish foods that will float, giving them a chance to eat and leave nothing to sink and dirty your system. Feed them up to three times per day to fatten them up nicely.

You'll buy your fish from any aquaponics fish provider and they'll be small fry when you get them. You'll need to have your system up and running for at least two weeks prior to adding fish and make sure that your pH levels are correct. You'll want between 5.5 and 6.5 to be optimal for both the fish and the plants.

Most varieties can be harvested as early as 1 year or wait until 2 years. If you keep them over the winter, you'll need to make sure that your temperatures won't dip too low to kill your fish. If you don't have the best climate for keeping fish through the winter, you should consider a faster-growing variety that you can harvest every 9 months. This will ensure that you don't lose any to excessive cold.

This is most often referred to as 'cleaning' fish. If you are working with scaled fish, you'll want to remove the scales. This can easily be done with a scaling knife. You simply run it across the sides of the fish and the scales will come right off. Hold the fish with your thumb in its mouth, making it easy to hold and flip from side to side as you remove the scales.

Use a sharp knife to cut a line around all the fins, then grab hold of them and pull them out with a pair of pliers. You may chop the tail fin off or leave it. Some people enjoy eating that if deep-fried.

Cut carefully from the throat area to the lower belly and around the vent, this will allow you access to the innards which can be scooped out in one motion. You'll be able to rinse the fish well under cold water and freeze it from this point. If you've got large enough fish to make it easy to fillet, you may use a filleting knife and carefully slice fillets from each side of the fish and discard the rest.

Some people like to clean their fish on top of a newspaper so that they can wrap the innards in the paper and toss it into a pile to burn the remnants. Nothing attracts bugs faster than fish guts.

Goats

If you are in a suburban area that allows small livestock, consider miniature goats. Goats are relatively hardy, easy to care for and make for good entertainment as they are very playful and intelligent.

Reasons to raise goats:

- They can be milked to provide you with
  - Milk
  - Butter
  - Yogurt
  - Soap and skincare products
- They can be butchered for meat
- They can be bred to sell for cash
- They will keep your lawn mowed and clear brush
- They have poop that makes an excellent fertilizer for your garden or can be fed to your aquaponics fish.

Miniature goats are adorable, and the most common varieties include:

1. Nigerian Dwarf
2. American Pygmy

Miniature goats are known for producing milk that is very high in butterfat content, more so than cow's milk. This makes excellent butter and cream products. You can even make yogurt and cheese from goat's milk. All from an animal that isn't any bigger than a cocker spaniel in size.

The market for skincare products and soaps made from goat's milk is an ever-increasing market as well. Your goats can lead to quite a side income and are a favorite amongst homesteaders. So popular, in fact, that purebred babies, known as kids, can fetch prices as high as $300 each shortly after weaning.

Additionally, goats that are milked directly following kidding will continue making milk for up to a year or more without being bred again. This isn't possible with other animals. It is also important to note the goats almost always give birth to two kids at a time, sometimes even having triplets. It's rare to only have one kid.

That means that if you breed your female, she could produce two babies each year, with a gestation period of only 150 days. Even with time off between pregnancies, she could produce 2 to 6 kids every 18 months at a sale price of $150 to $300 each, depending on their breed and parentage. In other words, investing in a good starter doe. She's a doe until she has kidded and then she is called a nanny goat.

Goat meat is very popular in every country in the world, aside from the US that prefers beef. There are many markets to sell goat meat and you can butcher one to two per year for your own freezer and easily put 75 to 100 pounds of meat in your freezer, even with miniature goats. Goat meat is red meat, unlike chicken and rabbit. This will add variety to your menu and be excellent for stew meat and roasts.

Goats are great with kids, especially the minis. They aren't big enough to do any damage and if they are polled (a process of burning their horns so they don't grow) then they are essentially harmless. They can be hard to keep in a fence, so you'll need sturdy fencing that is well-built and tall. Goats, even little ones, can jump fairly high and climb too. Hotwired fences are often used for goats for this reason.

You don't have to have a Billy goat to breed your female. Artificial insemination is common nowadays or taking your girl to meet a boy at someone's farm is also possible. Billy goats have a very pungent odor and aren't welcomed, even in rural areas that are zoned for livestock, you won't enjoy smelling a Billy goat and neither will your neighbors.

Butchering Your Goat

This is virtually the exact same process as butchering a rabbit. It's not difficult though it is a bit heavier to hang. Initial dispatching of the goat is most often done as a bullet to the head, followed by an immediate cut to the neck jugular vein. This will make things very swift. Meat processing facilities use a bolt gun to the head.

With the goat dispatched, you'll find it easiest to tie both legs separately and pull it into the air with a winch or using a rope over a dolly that you tie off to something. Once hanging, you'll start by slitting the throat and allowing it to bleed out as you cut the skin from the body.

Start cutting around the lower part of each leg until you have separated the skin and then cut straight down until both leg cuts meet in the middle of the abdomen.

Pull the skin down until you reach the genitals and cut round them. If this was a male, you may wish to pull the testicles out and save them. Many consider this a delicacy. You can cut the skin around the anal area and then pull downward again. This takes some muscle and if the skin begins to tear, stop to use a sharp knife to loosen the skin from the flesh before pulling again.

You'll be able to pull the skin all the way to the head, which you can cut off with a sharp knife or ax and continue pulling the skin down the front legs until you reach an area where you can simply cut the leg off and the skin with it.

Now, you want to go back to the abdomen and pull it out slightly and poke one finger-sized hole that you can slip your finger into and pull the skin out as you cut down and through the ribs, right to the neck. Now, reach inside, move the intestines to the side and remove the kidneys, liver, and heart so that you can save those. Let the intestines and so forth dangle, being careful not to cut into the intestines.

Cut the front legs off at the shoulder joints. Cut the joint by lifting the leg and move slowly as you work your way to the joint and then cut cleanly through the joint, removing each leg completely.

Now you've removed everything of importance below the intestines. Go to the top of your hanging carcass and turn it around so you can cut down both sides of the back. You can clearly see these backstrap muscles. Once you've cut the sides, make cuts across the top and bottom. Then you can pull your back straps from each side, you'll have two when you're finished.

Moving back to the other side now, you'll remove the hind limbs from the body, one at a time. Find the joint the same way as you did with the front legs, slowly cutting down to the joint, manipulating the limb to feel for the ball joint of the hip until you can see it and cut through it.

As you remove each leg, be careful that the carcass will shift and ultimately fall into your bucket below. Don't allow the intestines to break and splash onto any of your processed meat. Discard the rest of the bones.

You might be able to trim some cuts of meat from the ribs and you can do that by chopping away the anal area, still attached to the intestines. Once those are gone, you can trim more meat from bones that you can use as stew meat and process the rest of your pieces as you like.

Chapter 3
Curing and Smoking Meat to Preserve

Historically, curing meat was the best way to preserve it when there was no way to refrigerate foods. The refrigerator wasn't invented until 1834. Prior to that, foods were cured and smoked, pickled, and fermented.

The only way to load meat on board a ship, back in the days when Columbus was out discovering things, was to store salted pork in the belly of the ship. Sometimes, live chickens were kept so that the crew could have fresh meat at sea. Traveling back then was a process that took many months. Ships sailed, depending on the stars for directions and the ocean current to take them on their way.

The only way possible to take enough food for a large crew and passengers was to pack dry goods into the belly of the ship. This included bags of flour, rice, salt, crates with a few chickens sometimes, and root vegetables that could keep a long time.

Many trade ships stopped during their routes at familiar places to take on more supplies or go forth in hunting excursions that might produce meat. The first night, fresh meat would be enjoyed in a feast while the rest of the meat was salted or smoked. It takes only a day or two to smoke meat. Fruit was something that would not keep, and this is why so many sailors suffered from what was found to be a deficiency of Vitamin C.

Homesteaders on the prairies of the early 1800s didn't have a way to keep cold milk or fresh meat. Everything was consumed right away. Milk was used to make butter, milk, gravy and when it was drunk, it was warm and fresh.

900 years before the birth of Christ, salt was being used to cure meat in ancient Greece. The process hasn't changed a great deal since then. The salt trade was what initially began world trade and why salt was so important was because people needed it to cure meat and prevent it from spoiling.

Salted pork is still popular today in dishes such as baked beans or southern bean recipes. If you've ever had corned beef and cabbage, you've eaten salt-cured meat.

How to Salt Meat

The first thing you need is common table salt. Salt doesn't allow bacteria to grow and since bacteria are responsible for rotting food, anything that is salt-cured will not rot and therefore, not spoil.

The nitrates and nitrites in salt will turn the meat pink, thus explaining the bright pink of ham and corned beef. It doesn't take a lot of salt and it will be sometimes used in a solution known as a *brine*, or dry and mixed with other dry ingredients to create the flavor you prefer -- sometimes including sugar.

Salt is used in pickling for the same reasons, microorganisms and bacteria cannot survive in salt-laden environments. Botulism is a disease that is often fatal, spread in meat and other foods by bacteria. Salt creates an environment that doesn't allow botulism to grow.

1. Trim your meat so that there is very little fat. Getting a good cure is dependent upon the meat being dried well and fat is not conducive to the process. Fat will promote the cut going rancid because it cannot penetrate the salt.

2. Beef and pork are best for curing. Fish can be cured but isn't good meat to start on for you. If you're going to salt meat, you'll need to do it in winter, do it outdoors or have a refrigerator with plenty of space. Curing temperatures are between 30 and 50-degrees Fahrenheit, respectively. Jerky is best done as dehydrated and/or smoked.

Brining Method

1. Prepare a 14% solution of pickling salt. This is the salt that is used to make your pickles also, so having it on hand is always a great idea. To create different flavors that make it more palatable to you, make your solution 11% salt and the other 3% as brown sugar.

2. Your lean meat should be no more than an inch thick.

3. Soak it in the brine for only five minutes. This brine solution can be re-used during this one-day period only.

4. Place the soaked strips in a colander. Let them drain.

5. Hang your meat strips on a line. You can use hooks made from metal S-hooks. Some people just use heavy gauge wire stuck through the meat and used to hang it over a clothesline or over a board.

6. Drying takes 2 or 3 days in a dry, sunny environment. The best conditions are at 30% or lower humidity, with a flow of warm air that stays about the same temperature through the process.

7. Screened cages are recommended to keep insects away from your hanging meat. You also can use a smokehouse to dry the meat, in which you'll simply practice normal smoking procedures once the meat has been soaked. Place it in your smoker for up to 24 hours.

Dry Rub Curing

This method works well for large cuts such as ham shoulders, pork butts, and large roasts. You will, again, use pickling salt and rub it directly on the meat. Feel free to add special spices for flavor, things like cracked peppercorns, mustard seeds. You can even use fresh herbs from your garden. Rosemary will be delicious, but you should experiment to find the recipe that you like best. Rub your mixture of spices on first. Then rub on the salt.

You'll need approximately 1 1/2 cups of salt for each pound of meat but only apply half this time. Now hang your meat in a room having temperatures between 35- and 50-degrees F. In the winter you can hang it in a shed and cover it with mesh or muslin to keep insects off. Leave it like that, in temps between 30 and 50 degrees for 4-5 days. Then you'll take it down and run the last of that salt all over it. It will be ready after hanging another 5 days unless it is a bone-in cut. For meat containing bones, add two more days for a total of 7 days at each brining.

Now you will have a way to store meat when your freezer is totally full. Maybe you live off-grid and have no refrigeration? You can still enjoy meat. Salted meats will typically last up to 2 weeks without any refrigeration. Meats that are cured and smoked -- dried thoroughly in that process -- can last indefinitely. When you want to use them in stews or soups, they can be rehydrated in water prior to cooking.

Chapter 4
Composting in Your Backyard Garden

You may have heard about composting but had no idea just how scientific it can be. Composting is a process of breaking down organic matter and using it for fertilizing your garden. Compost is a key piece of the backyard homestead and you'll want to pay close attention to all that you can do with compost and the ways in which you create it.

This process takes some time but will help you to create amazing organic fertilizer for your gardens that will improve your soil over several seasons. You can turn the worst growing soil into rich, dark, organic soil that will grow anything.

Composting helps you to leave a smaller footprint on the planet as well. You're using your waste and scraps so that you are utilizing your waste in a way that benefits the planet. You are also growing food in a natural way that isn't introducing chemicals into the soil that will inevitably taint the water tables that find their way into human mouths one day.

There are repercussions to throwing things into landfills and creating more waste. Our goal should be to keep a healthy planet because we are the stewards of this amazing world. If we take care of it, as you'll find out by providing for your family from your backyard, you'll feel a sense of being at one with the planet and likely take your role as its keeper more seriously than you ever have before.

We are responsible for every little thing we do to this planet. Using every bit of waste that you can is a great way to ensure that you're doing your part. You're also getting your money's worth from the time and effort that you're putting in. There are many ways to utilize the waste you produce, even the waste from your animals.

Using their poop as fertilizer is just one way that you can manage their waste. You can also find some very creative ways that might have benefits you've never considered before. A little creativity can go a very long way.

If you are interested in doing any of the no-till gardening methods, then your compost can be used in all of them as well. It's essential to return nutrients back to the earth because every time you grow a new crop, it is pulling those nutrients back out again. It's a cycle that demands you to replace what you take and composting all of your scraps is a fast way to do so. It's timesaving, cost-effective, and reduces waste. A win for the garden, a win for the planet, and a win for you.

Twist on The Aquaponics

This method will combine the aquaponics system to include your vegetables, fish, *and* rabbits. The food your rabbits eat is largely vegetation. This makes excellent fish food. By digging a space for your pool to drop into, you can extend their season by keeping them warmer for longer in the ground.

By lowering your pond, you'll be able to access your plants easier and you can build your rabbit hutch so that it is over the edge of the water. Instead of having buckets underneath to catch your rabbit droppings, your fish will be eating a slow and steady diet all day long. You will have nothing to clean up from the rabbits, your fish are fed for free, and the poop is still fertilizing your crops in a roundabout way.

It's a system that is nearly on autopilot. You will need to make sure that you aren't providing too much food to the fish. This will result in some bottom junk that you'll have to use a pool vacuum to clean, so work to find just the right mix and it puts your aquaponics nearly on autopilot -- just keep feeding your rabbits good food.

Since rabbits are vegetarian, it's perfectly acceptable and healthy to feed the poo to your fish. Fish are also largely vegetarians though they will eat insects. Their job in your pond will be to eat mosquito eggs and stray insects that make their way into the water. You'll likely end up with some resident frogs as well. They're all good for the environment so don't worry.

Twist on Chickens

If you allow your chickens to free-range your backyard garden from time to time, they'll eat bugs and poop all over the garden areas, fertilizing things as they walk and cleaning your garden as they go. Turning them or goats loose in your garden after you've harvested all your fruits and vegetables is a great way to take care of the overgrown vegetation and they'll poop as they work, restoring it to the soil.

Be cautious to not let goats loose with berry bushes and trees because they will eat them down to stubs. Goats can be rented out to clear property for other people because they are so good at eating brush and weeds too. Goats can climb and they'll go right up in a tree and strip it bare, so be cautious with them. Chickens are a great choice though. Some people have good luck in turning them loose in gardens that aren't picked yet, but the plants are big enough not to be trampled. Supervision is key.

Letting chickens roam during the day should depend on where you live and how many predators are around. Foxes and coyotes both will nab an easy meal in broad daylight. You can hardly blame them, it's the circle of life. You simply need to ensure that your fencing and location is very secure if you're going to allow your hens to free-range by day.

A piece of PVC pipe slipped over a top wire all the way around your fence is a great way to deter foxes and coyotes from jumping over. They hit their front paws on the PVC, and it spins them right off. It's a cheap fix that often works. The higher your fence is, the better, but you will never 100% keep predators out, so you must be vigilant with your girls.

Chapter 5
Biofuel from Compost

The first thing you want to know is, "what is biofuel?" Let's discuss what it is and why you'd want it. It's great stuff and after you read this section, there is no doubt in our minds that you will also agree. In fact, you'll be shopping for a biodigester in short order, we're sure.

Biofuel is a type of fuel created from methane gas formed from compost with manure added to it and is kept in an anaerobic environment. In this case, it's kept in a sealed container that is specifically made to do this. Being anaerobic simply means that it is an oxygen-free environment.

As the organic matter, mixed with manure from your animals decomposes, it creates a gas that can be siphoned from the container and be used as fuel. In fact, it will operate most generators, small engines such as lawnmowers and weed eaters. It will even work in many cars and trucks. It's not hard to do this because you just have to add your refuse and then reap the rewards after the biodigester does the work.

The matter that forms in the bottom of the container can be harvested as some of the best fertilizer on the planet. It's so good that you could bag it and sell it. You can't sell the biofuel and you wouldn't probably want to anyway. You'll only make small amounts, but it can be enough to run your lawnmower all summer. It depends on how much organic matter and manure that you are pumping into your biodigester - the name of the composter that makes biofuel.

You can expect to spend anywhere from $700 to $1500 on a small backyard biodigester but you can also make your own. There are many plans on the internet. China, for example, has millions of them in use throughout the countryside. Millions of them help create fuel to provide generator use to those who are far away from the electrical power grid.

Those who live in rural China are far removed from a quick trip to the gas station for fuel to create their power or operate a small car or machine on their farm. As of 2005, more than 20 million households in China were powered by biofuels. The technology is available for you to have the same benefits of use. After the initial investment, biofuel pays you back in many ways with many potential uses. Why not take advantage of fuel you can make from your own compost?

Chapter 6
Foraging for Foods in Your Area

One method of gathering food that is often overlooked is foraging for what is growing in the wild. There are many foods that grow all around you and can be found in both urban and rural settings. These can give you a wonderful fresh salad any day of the week during growing seasons.

Wild Foods

Wild greens and foods of all types are found in all zones across the US. Some are more popular in specific areas of the country. These greens are best when picked young so they are most tender and should be washed very well to ensure that they are very clean. Many edible plants are simply written off as weeds today, but our forefathers used them for food and for medicinal purposes.

There are also mushrooms, which should also be picked early in their season. Morels and puffballs are both examples of these. They are common in many parts of the United States and delicious, free food. In fact, they may be more delicious because they are free.

This information is certainly not enough to make you an expert and you'll come across a disclaimer below because of that. It takes years to learn which plants are safe and which ones you have to prepare in specific ways.

Be safe always, take some courses, do some hiking with park rangers when you can, and learn from others who have collected wild foods for many years. There's a lot to be found out there and if you teach your children these things, you'll never have to worry about them being lost in the woods because they'll know what to eat to survive.

Sloe - This is the fruit of the blackthorn. They may resemble a blueberry at first glance. They start green and turn dark blue as they ripen. They are sour to pungent and quite edible. Very common in Europe where they were first used in Sloe Gin. This is also sometimes known as a wild plum and grows from the deciduous tree known as the Blackthorn. It's most often found in the sunshine.

Dandelion Greens - The dandelion is considered a weed, but it is actually one of the best sources of wild food that you can find. The tender young greens can be sautéed with some butter and incorporated into salads raw. The delicate yellow petals are edible as well. The roots can be dried and added to teas, considered a kidney cleansing agent. Deep-fry the entire head of the flower after dusting in flour for a delicious snack in the form of fried mushrooms.

Chickweed - This is a commonly found weed that has been used for medicinal purposes for centuries and is also a fresh addition to salads. It has fine hairs on the stems and the petals are white and grow in pairs of two for a total of ten petals.

Wild Asparagus - Identical to the garden variety, this can often be found growing alongside the road in places like Illinois in the spring. Asparagus is delicious in butter, sautéed, or steamed. It can be eaten raw alone or in salads as well.

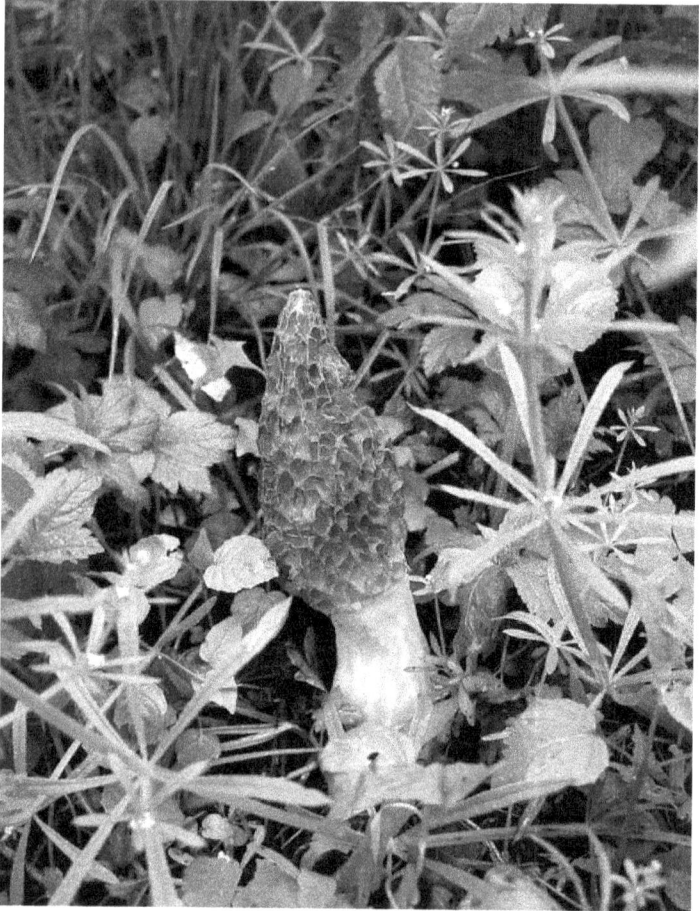

Morel Mushrooms - This is a variety that grows in shady, moist, wooded areas. The morel is worth upwards of $25 per pound and mushroom hunters will not disclose the locations that they've found morels because of the worth and the competition for selling them. Restaurants will buy these from you if you find them, but they are also delicious to eat at home. They are distinctive and easy to spot for the novice mushroom hunter. Soak them in saltwater prior to eating them to ensure that they are clean and bug-free.

Chicory - This is a flower in the same family as dandelions. It's bright blue and a tasty addition to salads. It can be found all across the US, particularly in the southern states, as far south as Florida. The petals are dried, grounded, and added to many things.

Broadleaf Plantain - This plant is very similar to spinach in flavor. It's best if eaten when the leaves are young and still tender. It's a common plant found in many states across the US. People regularly spray to kill it, but the plant is edible and is a great addition to salads or in any application where you might use spinach. Cook it or eat it raw in salads.

Crab Apples - These are not native to America, but they've been here a long time. They were brought from Europe and you can find them in scrubland areas. Harvest them when the apples turn red or begin falling from the vines, usually in late summer.

Kudzu - This is considered an invasive species and is very hard to get rid of. This makes it easy to find in many places. It grows like a jungle weed in states such as Georgia, strangling trees from the light of day. When you are driving down the interstate in Georgia, the thick vines you see to each side is kudzu. It's a vine that grows all year, but the broad ivy-like leaves fall off in the cold seasons. This entire plant is actually edible, even the small flowers.

Mullein - This is a very popular plant amongst foragers. It can be eaten, dried, and used in tea. Even the flowers can be eaten. This is sometimes referred to as the 'toilet paper' plant by hikers because the leaves are soft and broad. This plant has been known to grow higher than six feet tall.

Red Clover - An edible plant that has tasty flowers as well. The leaves and the flower are the edible bits that you'll want to consider. It's a wonderful addition to any salad for both color and flavor.

Sheep Sorrel - This plant grows in acidic sandy soils, in humid grassland environments. It has a taste of lemon and can be eaten raw if you like.

Bee Balm - This wild plant has a lovely scent and the flowering heads produce edible petals. They may be a delicious snack if you are on the trail or foraging for additions to your salad.

Wild Leeks - These grow in many places in the US. They can be distinguished by their long dark, grass like leaves that grow straight up in the air and you'll likely smell them before you see them. The onion smell is quite strong. Gather these and take them home to add to any dish that calls for green onions or scallions. They may be purplish in color at their base and you shouldn't eat a lot of them in one sitting.

Wild Blackberries - These are very much the same as their garden variety cousins. They are often found growing in sandy, well-drained soil. They can grow in abundance if you are lucky enough to find them. Do be careful of the thorns. Make them into a jelly that you can preserve by canning.

Wild Blueberries - These aren't planted, they grow from shoots in the wild. They can be slightly tarter than their cousins but are nonetheless delicious. Enjoy them in muffins, pancakes, as fresh salad additions, or simply eat a bowl of them. These can be cooked into jams or syrups and preserved.

As you can see, there are many types of foods that you can gather via foraging. Remember to make sure that you aren't trespassing on private property. When you see a tree with fruit on the ground under it, knock on a door and ask if you can pick some fruit. Many people will tell you to help yourself if the fruit is falling and they've not harvested it themselves.

Hazelnuts

Nuts can often be gathered from taking a walk in the woods. Walnuts, pecans, hazelnuts, and more grow in abundance at specific times of the year. These need to be gathered before they rot on the ground. Some will require that you dry them in the oven. Look up the directions for how to properly prepare these things.

Recipes for Foraged Fruits and Nuts

Now that you've found some wild foods, what do you do with it? There are many recipes to use your finds and thanks to the internet, they're not overly difficult to find, but here are a few that are wonderful favorites that have been lost to time.

Blackberry Meat Sauce

1 pound of blackberries
3 Tablespoons of confectioner's sugar
1 Tablespoon of lemon juice

Directions

Mix all three ingredients in a saucepan, bring to a boil. Then reduce heat to a simmer for 10 minutes. Allow sauce to cool, then add to a blender and blend until very well blended. Pour through a strainer to remove seeds. Place in a glass, sterilized jar and refrigerate. It will keep up to 90 days and is delicious served over pork, duck, or turkey.

Crab Apple Jelly

2 pounds of crab apples
Confectioner's sugar
2 cloves

Directions

Chop the apples and place them in a pan along with the cloves. Add water until just covered. Bring to a boil and simmer until the apples are soft and pulpy. Skim away any foam that forms on top of the water as they cook and discard. Place the apple pulp in a jelly bag overnight in a saucepan to drain.

The next day, toss the pulp into your compost pile. Measure the juice and add 1 pound of sugar for each pint of juice you've got. Heat this mixture slowly. Once the sugar is dissolved bring to a rapid boil for 20 full minutes. Pour into sterilized jars. Seal while still warm.

Sloe Gin

Wash the sloes thoroughly under cold running water and then pat them dry. Use a sharp knife or prongs of a fork to prick each one of them. Ensure that you only use ripened fruits. Place them in a pre-sterilized, wide-mouthed glass jar.

Weigh how many sloes it takes to half-fill your jar. Then add them to the jar along with ½ their weight in sugar. Top the jar off with gin. Add 1-2 drops of almond essence or extract. Drop a whole cinnamon stick into the jar, seal it, and shake.

Store this jar in a cool, dark cupboard. Give it a shake each day for 2 to 3 weeks. After the last shake, allow the jar to sit untouched for 3 to 4 months. It will turn into a red-colored juice that you can then pour into a bottle and discard the other ingredients. You now have sloe gin.

Flour from Acorns

Gather your acorns as soon as they fall before they have a chance to rot. Transfer them to a baking sheet and roast them for 20 to 30 minutes at 250-degrees. This is to kill any potential parasites and dry them thoroughly.

When they've cooled, crack them and remove the meaty centers to a bowl. Once you've completed this task, add the centers to a muslin bag, and soak them in water for TWO WEEKS. Change the water each day to ensure that you remove as much of the tannins from the nuts as possible. These are bitter in acorns.

Sun-dry your nut meat at the end of the two weeks, just until they are free of moisture. Grind them immediately. You can use this flour in place of any wheat flour recipe or convert recipes to use them as if they are wheat flour. You can make pasta, dumplings, pancakes, and more. Store the excess flour in a sterilized glass jar with a sealed lid. You may also store it in the freezer.

Candied Violets

Pick your violets by leaving a little stem to hold onto. Wash them gently so you don't remove the petals. Allow them to dry on a paper towel or kitchen towel until they are thoroughly dry.

In a saucepan, heat ½ cup of water, 1 cup of sugar, ½ teaspoon of almond extract. When sugar is totally dissolved, use tweezers to carefully dip the flowers into the hot liquid. Dust them with sugar until they are totally coated. Use them to decorate cakes, cookies, top a salad, or just eat them as is.

A Word of Warning About Foraging

Do not eat anything unless you are 100% certain of what you have. There are poisonous plants out there that look like edible ones. Mushrooms are specifically deadly if you eat one that is poisonous.

Do not take a chance. Learn all that you can from someone who is knowledgeable about plants. There are often courses available through local clubs and park districts. There are apps that can be downloaded to your phone that will help identify plants but don't rely on these for accuracy. They've been known to be wrong.

Also, never trespass on private property. Property owners are often happy to let you pick some blackberries, wild grape leaves, and so forth. Ask permission first. It is illegal to remove plant life from state parks and national parks. Unless they specifically give you permission to pick berries, you are not allowed to do so. They are there for wildlife.

Never pick all of what you find. It's important to preserve the next season's crop, so allow some berries to be consumed by wildlife, who will transplant the seeds in their stool and ensure a new crop next year. Too many times, people don't understand this and pick every single berry they can find. This decimates nature.

Many state parks do wildlife walks in the spring and summer months. No one is able to teach you more than the wildlife conservation officers of your local park district and tour guides. They'll answer your questions and teach you about your local plants and wildlife so that you can learn from experts. Often, these trail hikes with rangers are free or very low cost.

Chapter 7
Building the Things You Need

There is no doubt that you're going to need and want some things to help you hold livestock, keep your gardens, and process your foods. Many things can be built yourself and some can even be made from junk that you already have laying around, or that you can salvage for free or very little.

Trade some eggs for what you need with a neighbor. You'd be surprised how many people still enjoy a good barter. Bartering was popular long before everyone had money. In fact, bartering was just as good as cash and we may very well see a time when we rely on this again. He who has the most that can be bartered may indeed be worth more than others.

The Chicken Coop

A chicken coop needs to be structurally capable of keeping predators out. Don't think that because you live in the city or the suburbs that you don't have wildlife predators, you do. You've perhaps not seen them, but they are there. Foxes roam downtown Denver at night and coyotes howl in the outskirts of the suburbs, waiting to prowl the shadows after you are in bed. Raccoons are everywhere, as are snakes and possum.

Chicken wire does not make the best wire for a chick coop. Use hardware cloth. It's a heavy gauge and rigid. The holes are too small for a predator like a raccoon to reach through and grab a hen. Even a coyote can reach through a chicken wire fence and the force with which they will yank your chicken through will break the fence and kill your chickens. Predators will return to easy targets too.

Pallets can be found very cheaply if not free in garbage piles. You can pull them apart to use the boards or you can use the entire pallet as a section of wall for anything you like

With hen houses, you want them to have an indoor coop that is a few feet off the ground, with a ramp they can walk up. They like to roost up high at night. Inside their coop, you'll want to have bedding material and some nesting boxes where they lay their eggs.

Build a door that allows you to access the nesting boxes from the outside and gathering eggs will be much easier. You will also need an entry door that allows you to clean inside. Change their bedding regularly to keep it clean and fresh.

Inside their yard, give them more than one location for water that is clean and fresh. Waterers for chickens are designed to keep the water clean but you'll need to check it and change it regularly. Inside your coop, add some places for your hens to roost on shelves or 'chicken trees' like a cat perch.

The bottom line is that if you keep the wire with tiny holes, like hardware cloth, snakes and predators cannot access your run. Digging a trench and extending metal sheets or your hardware cloth down into the ground for 12 to 18 inches will ensure that predators who dig (coyotes, fox, raccoons, and possum) can't get into your run when you aren't looking. A hungry fox will strike during the light of day.

Cover your run because chicks and chickens look very good to a hawk or an owl. They will swoop down and snatch your birds before you can say 'stop' and it will be too late. Hardware cloth over the top will ensure that nothing gets in. If cost is a factor, an old net from a trampoline works great. If you can get your hands on the whole trampoline, you can fashion an entire coop and run from one.

Goat Run-In Shelter

Goats don't need a completely enclosed shed and will develop respiratory infections if you keep them in a closed barn. They need a three-sided structure or open building that allows air to flow through at all times. Give them a roof to hide from the rain and plenty of bedding to flop down in and you'll find your goats happily napping in a structure made with pallet sides and a tin roof. They don't care what it looks like, as long as they can stay dry. Goats don't appreciate getting wet.

A quick run-in shed is something that is one of the easiest things you'll ever put together because you can use the pallets exactly as they are. Use screws to attach them together into three walls and lay a sheet of galvanized roofing material over the top and attach it. You're done.

Garden Fence

For this, you can get very creative. Any fencing material will do, including things that might not normally be fencing. Make sure that you use something that won't allow small rodents or rabbits in at the bottom. If you have deer in your area, you'll want a fence that is 6 feet tall to keep them out of your garden.

One or two deer can devastate your garden overnight, leaving you with nothing to show for all your hard work. Make sure you put something up around your garden and understand that four feet might not be tall enough if you have deer.

You can always add rope around the top, or wire, and string some things on it that will also keep birds away. Tin pie plates work well sometimes. Birds will steal berries, and small vegetables when they are hungry. They'll also take your seeds before they've even sprouted.

A Smoker

This is a wonderful way to preserve meat, especially pork and beef. You can fashion your own smoker by scavenging bricks, stones, an old wood-burning stove for parts, etc. It doesn't take a whole lot to build a smokehouse.

A wood stove that you can set up outside but run the pipe inside of a shed that you build to hang meats inside of and slowly smoke is a fantastic way to preserve meat for longer and it tastes amazing. Smoked BBQ is some of the best meat on the planet and it's a bit of food art. Those who smoke their meat are picky about the type of wood they smoke because it changes the flavor of the meat.

A little bit of research will show you a hundred different ways that you can fashion a smokehouse from salvaged materials. Add your own unique ideas and make it your own design. Cement blocks for the base of your building and the woodshed on top of them to smoke your meat in is a simple concept. It's a way of cooking meat that has been used for ages.

Some people like to build their smokehouse from cedar. Cedar is a wood that deters bugs and it won't rot. It also has a pleasant smell which seems to add to the smokehouse experience.

Inside your smoker, you'll want rods to hang meat on or hooks that can hang from above. The more you can smoke at one time, the more you'll be able to put away and store it at one time. You'll want enough space to smoke a whole goat when you butcher it so that you can process it all at once.

One of the quickest and easiest ways to create a smoker is to use an old refrigerator. The older, the better. Place a sheet of metal across the bottom interior of the fridge. Place a can of wood chips in the bottom or a small grill that you can load with wood chips. At the top of the fridge, cut three or four holes on each side for vents. Use sheets of hardware cloth for racks and smoke your meat the same day. An old stove will also work as a small smoker.

Fish, especially trout or salmon, are wonderful when smoked, so the smoker will get a good workout from your homestead livestock butchering.

Cold Frames

These will be for keeping your plants and soil warm for several weeks longer than in a typical growing season. They'll also allow you to get plants outside a few weeks earlier than typical.

A cold frame can be fashioned from bricks, cinder blocks, or built from lumber and lined with rigid foam board insulation. For the top, a piece of cut plexiglass will work, framed in a wood frame, or just find an old window that someone is getting rid of. When someone is getting rid of old windows, grab them. You'll find fifty different ways to use them and could even fashion your own greenhouse from them.

Cold frames don't have to be permanent but also can be if you've got the perfect spot for them. They should always be in a southward facing position, in the sunlight where they can gain maximum heat. Making your tops removable will allow you to cool them when they are getting overly hot. Adding a screen to the top can turn them into beds that keep moths and bugs away while allowing fresh air and rain.

Compost Bins

Again, a fan of pallets, you can fashion a three-section, open composting container that allows you to just dump into one section, move to the next section when that one is full, and rotate your compost in and out as it is ready. By having more than one section, you'll always have compost that is ready to use

If you don't want to use pallets, you can use other materials, including old garbage cans, provided that you make sure they can drain and let air circulate. The compost should be moist but not wet. You don't want it to be rained on if you can help it. You also want to allow excess moisture to escape. Compost shouldn't stink. If it stinks, it is too wet. It's the moisture that holds the smells in. Composting toilets work on this same dry composting theory.

If you've got cement blocks or bricks, you can also stack them into three-sided structures that you place compost inside of. As an added benefit, place your rain barrels over the top of them to keep the compost from getting too wet and get the bonus of compost that keeps your rainwater warm enough to not freeze in the winter months.

Locate your compost bins between your manure gathering stations (chicken coop, goat shed, rabbit hutch) and the garden where you'll be using the compost when finished. This means you'll carry it only as far as necessary.

Raised Garden Beds

Virtually anything can be recycled into raised bed gardens. Kiddie swimming pools work great. Add some holes for drainage and you're in business. If you want something that looks nice, you can use cement blocks and landscaping stones to make yourself a very nice raised bed without ever driving the first nail.

Find a building that has burned and left a mess. Often, owners will allow you to take what you want after they've gotten their insurance check. Look for used materials on Craigslist and don't forget to check the free section. A lot of people have materials that are there for the taking if you are willing to remove them or take something apart to get them. It's shocking what some people will give away for free, frankly.

An old boat or canoe can make a wonderful raised garden bed when the boat no longer holds water. For garden purposes, that's perfect. It also makes it worthless to someone who wants a canoe so you can probably get it for free or very cheap just to get it out of someone's yard or garage.

Think outside the box to create your whimsical garden or your English garden. It's all in your grasp if you keep your eyes open at all times.

Climbing Cages

Some plants need something to climb on. It doesn't have to be tomato cages. Some rope strung back and forth through two pieces of wood, stuck in the ground will work for a lot of plants to climb on.

Creating a wood frame from some free pallet wood and then wrapping some rope or wire through the slats can be made to look whimsical and work wonderfully well to support plants.

An old ladder missing a bottom rung can be buried into the ground deep enough to stand straight as a support for beans to climb toward the skies or a trellis for ivy, you decide. Just don't let the ladder go to waste.

Boards framed into an a-frame with cross slats are a wonderful climbing frame and you can actually prop two wood pallets together and make this in ten minutes with a drill and some screws.

Potting Bench

If you are doing container gardens, you'll really appreciate having a potting bench where you can do your work and have a countertop to use. Old kitchen cabinets can be used, or you can build yourself a nice counter with some 2x4 lumber or pallets.

A top can be cut from plywood or using an old door or piece of sheet metal. Shelves, drawers, or storage space for your tools will be a relief to have everything within your reach when you need it. You'll be able to pot faster and easier while saving your back.

A Simple Hoop House Greenhouse

This is a welcome addition to any garden. You can extend your growing season and protect young plants from rabbits, squirrels, and other problems, such as weather. You'll be able to keep them warmer, out of the damaging winds of summer storms, and the hail that can decimate a garden.

All it takes is some PVC, a few rolls of plastic film that can be purchased online or even at a Walmart. You'll need a little lumber, including one long 2x4 that can be used at the top as a ridge board. You'll drill holes the size of your PVC pipe, and slip that through, then secure the PVC into the ground on either side.

You can do this by securing them to 1x4 slats on the ground, using brackets. This adds some weight to the bottom, keeping it from blowing over easily. Those bottom slats can also be secured to stakes that are driven into the ground as anchors to hold it all in place.

The ridge board will hold each rib of PVC at equal distances so you can wrap the entire structure in plastic. Frame a door at one or both ends and use any old scavenged door as your entry.

Willow Fence

This is an English form of fencing by starting with willow tree limbs, stuck into the ground at an angle. When you've done all of your pieces in the row, you go back to the beginning and weave new pieces across those pieces.

You can get very creative with the pattern and how neatly or wildly you weave your willow cuttings. With luck, those pieces that you've started in the ground will take root and it will begin to grow into a living fence that will keep small creatures out while providing a natural place for birds to nest. It's a lovely addition to any garden when you need some division between sections or something pretty to look at.

Cardboard Box Dehydrator

A simple and easy way to dehydrate some of your fruits or veggies by means of a cardboard box couldn't possibly be simpler, or cheaper.

An ideal temperature range for dehydrating is between 125-degrees and 145-degrees. All you need is a large box, a thermometer, a light fixture for a 100-watt light bulb, and some duct tape.

For the inside of the box, you'll want something that you can use to fashion a rack to dry veggies on. A piece of hardware cloth could work, or a baker's cooling rack will work just fine. You can prop it on four upside-down cups in each corner or come up with anything that works well for you.

You will use the tape to secure your box to the counter. Add your rack inside and your lightbulb in the fixture. Drop your thermometer into your rack to hold it in the center and close the box, taping it shut. Give it an hour and check to see what temperature you're at. If it is too hot, use a lower watt bulb, such as a 75-watt bulb. Try again and when you are holding in the perfect range, start with something easy to dry.

Apples, peaches, and cranberries tend to be relatively easy. Layer them on your rack and leave them for about 5 hours and check them. Many fruits can take up to 8 hours. Enjoy your fruits and use this method as often as you like, saving much of your harvest as dehydrated goodies that can be frozen or stored in jars.

Nesting Boxes

Chickens need nesting boxes and too often they cost a fortune to purchase the fancy, nicely constructed boxes when your hens will lay eggs all over the place. It's true, once you've got chickens, you'll learn that you'll have to check every corner for eggs that some sneaky hens will hide from you. Some will lay them outside.

Finding materials that make great nesting places is ideal. Old tires have been used with a lot of luck by some farmers. Fill them with straw bedding and keep them clean and they are just the right size for a hen to snuggle down into and lay an egg or two each day.

Five-gallon buckets make good laying boxes, but they also cost as much as $10 each. If you've got a cat, or know anyone who does, find out if they use the plastic buckets of cat litter. Those buckets are perfect because they are flat on each side and the lids have a lip that will keep the straw inside without falling out. The lid can simply be left open like a little step or cut off at the lip.

There are many creative things to use as hen laying boxes and you can search ideas on social media and spend hours looking at the wonderful ideas, most of which are free.

Rabbit Hutches

A rabbit hutch, like a hen house, needs to be very solidly built to keep would-be predators out. The bottom of each rabbit's cage will need hardware cloth to allow their droppings to fall through to buckets below, which you can save for the compost pile and garden fertilizer.

Use heavy wood for your hutch, such as solid 2x4s for the frame and 4x4s for the legs. The carpentry skills needed for this type of building are moderate. If you can operate a Skilsaw, read a tape measure, and be careful to measure twice before you make cuts, then you can manage a rabbit hutch.

If you are completely devoid of carpentry skills, there are kits that you can purchase for $100 and up. You may cruise local ads and Craigslist and find one that is for sale that was built solidly by someone who no longer keeps rabbits.

Rabbits can be escape artists and they are also easy prey for predators who will try every which way to get into your hutches. Make sure that doors are secured in ways that a crafty raccoon cannot unlock. Rabbit kits are tiny, tasty treats to many other animals.

Chicken Tractor

This is a mobile chicken pen that can be slid, wheeled, or pulled across your yard. It's a wonderful idea to allow chickens to patrol for bugs in your entire yard while keeping places safe from them at the same time. Spreading them around will also spread fertilizer at the same time, by way of chicken poop.

To build a simple chicken tractor, use lightweight wood to frame a simple pen with chicken wire to house them just during the day and under your supervision. Attaching old mower tires at one end and a hitch or handle that can hook to a riding mower will make it simple to tow it to the location for the day and put your hens inside. At nighttime, you can tow it back to their yard, open the door and let them run home to roost.

Chickens are incredible at keeping bug populations down. They eat ticks like crazy and will be worth their feed in just the bugs they eat. They'll actually eat a lot less feed when they are feasting on grasshoppers and crickets too.

Remember to have water and a shaded area in this tractor for them too!

Chapter 8
Your Backyard Homestead Products

When you are working to provide food for sustainability from your backyard, by following the advice contained herein, you'll be able to provide the majority of your family's necessary food supply. In fact, you'll have the ingredients to make things that you hadn't thought about.

If you are able to forage for some additional things, like hazelnuts, you can make your own flour in a pinch. Making flour, having eggs, and adding some additions like salt, allow you to make your own pasta, pancakes, biscuits, and more.

You've got milk covered, the meat of all types covered, all the vegetables and fruits that you could possibly need. You've learned how to dry your foods, can them, pickle them, ferment them, smoke them, and make jerky of all types. You've even learned how to make a biofuel that can offer you fuel for your generator when power is down, or you need gas in the mower. What else could you possibly need?

Soap? You can make that with goat milk. Cleaning supplies? Lemons juiced from your potted lemon tree will cover this and make excellent glass cleaner too.

The point is that if you are creative, you can utilize the by-products from your backyard homestead to create a plethora of items that you'd not even thought about. Many people think that they don't have the time to do all of these things, but let's look at it in a different way.

Most people go to jobs to earn money to not just pay bills but to purchase food. In fact, food takes approximately 20%. Poorer households spend as much as 35% of their income on food while wealthier households spend roughly 8% of their income on food. Most people considering growing their own food will fall at the lower end of the income scale and are spending a higher percentage of income on food.

Imagine saving thousands of dollars per year on food.

What could you do with that extra money? Would you pay your house off sooner? Pay off the credit card debt that keeps you awake at night? Pay for your kids' braces in cash perhaps? Let's face it, that is a ton of money and one of the things you could do in time is to work less. Imagine not needing a second parent to work because you've no longer got to purchase food and cough up a gigantic portion of your income.

The icing on the cake is that there are a hundred ways in which you can actually earn income from your homestead. Turn it into a side job that keeps you home, making more than when you were schlepping out to a job, spending money on gasoline, using hours each week in a commute.

You're spending precious time to sit in a car that you had to have that job to pay for, and you needed it to keep that job. What a ridiculous circle we find ourselves caught up in!

A backyard homestead can help to give you back a lot of your budget, but more than that, it can give you precious *time* back. Your time is invaluable, but we trade our time for a weekly paycheck that barely meets the expenses and gives us far less satisfaction than working for ourselves can.

Chapter 9
Anyone Can Build a Backyard Homestead

The bottom line is that you can start small and grow big. You can use space up into the air or spread out to take more acreage. You can grow food in pots, containers that you repurpose from anything that will hold soil -- or water.

Plants will grow anywhere that there is sunlight and water to provide them with hydration. You can provide nutrients to water or to the soil. Caring for them isn't incredibly difficult though there will always be challenges. Gardening is a process. It's a mindset and a way to become healthy, mentally, spiritually, and physically.

Gardening can be challenging to your body, but it can be adapted to suit someone in a wheelchair or someone with a back problem. Adjusting the height by planting only raised bed gardens will allow someone who is in a wheelchair to care for plants each day, plant them, and collect them as they grow.

If you are an apartment dweller with only a small patio or grassy area outside your sliding glass doors, you can plant a garden in pots and containers. Plant things as high as you can reach, love them, and care for them each day and they will grow.

As you learn the joy of creating your own food and making meals with the things that you've created and cared for, you'll be filled with a sense of connection to the planet and the world all around you.

It's a spiritual thing for many people. It's also hard to not stop to think about your grandparents and the ones who came before them.

Gardens aren't just about the roots in the ground, after all. They are your connection to your own history and the way that humans have lived for thousands of years, surviving without the luxuries that we have become used to in the lives we lead today.

Technology has permeated every facet of our lives and as much as it is convenient, it's also draining. Sitting in an office, staring at a computer all day is hardly enjoyable after several hours per week, over the course of years. Working in cubicles surrounded by stale air being filtered into the room through ventilation systems that are likely filled with spores and dust for lack of cleaning … that's no way to enjoy life, is it?

Gardening gives your body and your soul a break from the daily doldrums that we have fallen into, like zombies on our way to work each morning, a cup of coffee in one hand, and the other honking at the guy who cut you off in traffic.

In your garden, you can leave that world behind and all the stress that goes with it. Get your hands dirty. Run around barefoot if you want to. Talk to your plants and let them take that stress away from you as they give you fresh oxygen in return. Nothing can beat the feeling of being surrounded by a garden filled with life-giving food. It's a truly symbiotic relationship.

Create a space to sit and close your eyes, just feeling at peace in your garden and enjoying the sunshine with them for a few minutes each day. You'll find that these garden meditations become your favorite part of the day, and you feel calmer the rest of the day.

As you eat fresh foods, you'll feel more energy because you are eating freshly picked foods that are filled with the energy of life. They are thriving with energy and nutrition and when you consume that, it's an energy that sustains your own body. It's fresh fuel, free of preservatives, hormones, wax coatings, herbicides, and insecticides.

The best part? *You created it.* It's not unusual for people to refer to their plants as their babies. You do definitely form a relationship with your garden. Sometimes, it's your nemesis when there is a problem and you've run out of ideas for how to fix it. That's when you turn to your friends on social media plant and gardening groups.

They will understand your anxiety and frustration with why your tomatoes aren't doing well. They'll remind you of how to get back to the basics and use some Epsom salt to encourage the soil to be proper for them to grow.

In time, you'll find yourself helping someone else. Some springs you'll have so many seedlings sprout that you'll be looking for a neighbor that you can give them to and the circle of life is now shared with them.

The more you do, the more money you'll save on your food budget. The more you will respect the food chain as well. You will have independence that cannot be described but you'll know the feeling once you have it.

One day, you might even feel bold enough to walk away from the job that drains you and find that you can earn money from your backyard homestead, by teaching others and selling your wares. In times like these, the time is good to be more self-reliant and the best time to start anything is always now.

www.ingramcontent.com/pod-product-compliance
Lightning Source LLC
Chambersburg PA
CBHW071425210326
41597CB00020B/3662